Manual of Water Supply Practices

M61

# Desalination of Seawater

Second Edition



American
Water Works
Association

Manual of Water Supply Practices—M61, Second Edition

# Desalination of Seawater



**Disclaimer**
The authors, contributors, editors, and publisher do not assume responsibility for the validity of the content or any consequences of their use. In no event will AWWA be liable for direct, indirect, special, incidental, or consequential damages arising out of the use of information presented in this book. In particular, AWWA will not be responsible for any costs, including, but not limited to, those incurred as a result of lost revenue. In no event shall AWWA's liability exceed the amount paid for the purchase of this book.

Sr. Editorial Manager, Books: Suzanne Snyder
Manager – Manuals: Molly Beach
Manager – Publishing Operations: Gillian Wink
Specialist – Copyright and Permissions: Peggy Tyler
Director of Publishing: John Fedor
Technical Editor: Anne Lanctôt
Technical Illustrations: Michael Labruyere
Cover Image: antishock/Shutterstock.com
Production: Innodata

**Library of Congress Cataloging-in-Publication Data**
CIP data pending from the Library of Congress.

ISBN 978-1-64717-260-2 ISBN, electronic 978-1-61300-783-9

http://doi.org/10.12999/AWWA.M61ed2



American Water Works Association

American Water Works Association
6666 West Quincy Avenue
Denver, CO 80235-3098
awwa.org

# Contents

# Figures

# Tables

# Acknowledgments

The AWWA Technical & Educational Council, the Water Resources Sustainability Division, and the Water Desalting Committee gratefully acknowledge the contributions made by those volunteers who drafted, edited, and provided the significant and critical commentary essential to updating M61.

**Chair**

*Greg Wetterau*, CDM Smith, Rancho Cucamonga, Calif.

**Quality Control Committee**

*Angie Brown*, AWWA, Denver, Colo.
*Casey Rodriguez*, AWWA, Denver, Colo.
*John Thompson*, Desalination Company of Trinidad & Tobago Ltd., Point Lisas, Trinidad
*John Tonner*, American Membrane Technology Association, Englewood, Fla.
*Greg Wetterau*, CDM Smith, Rancho Cucamonga, Calif.

**Contributors**

*Giancarlo Barassi*, Aquatech, Tampa, Fla.
*Mark Donovan*, GHD, Inc., Irvine, Calif.
*Keith Downer*, Seven Seas Water, Tampa, Fla.
*Val S. Frenkel*, TYLin International, San Francisco, Calif.
*Mo Malki*, American Water Chemicals, Inc., Plant City, Fla.
*Laila Martin*, Tampa Bay Water, Tampa, Fla.
*Erin Messner*, Colorado Water Congress, Denver, Colo.
*Chris Morss*, US Water Services Corporation, Truckee, Calif.
*Rubén Muñoz*, CDM Smith, Dallas, Tex.
*Chandra Mysore*, CHA Solutions, Atlanta, Ga.
*Christine Owen*, Hazen and Sawyer, Tampa, Fla.
*Christian Sanders*, Veolia Water, Sydney, Australia
*Say Sengsouvanna*, Iv3 Aqua Corporation, Lakeland, Fla.
*Kevin Thomas*, Kimley-Horn and Associates, Inc., Riverside, Calif.
*John Thompson*, Desalination Company of Trinidad & Tobago Ltd., Point Lisas, Trinidad
*Cecile Toupiol*, CDM Smith, Amman, Jordan
*Rajendra Vaidya*, PCL Construction, Inc., Tampa, Fla.
*Srinivas (Vasu) Veerapaneni*, Black & Veatch, Overland Park, Kan.
*Nikolay Voutchkov*, NEOM, Sharma, Saudi Arabia
*Steve Wait*, GHD, Inc., Santa Cruz, Calif.
*Peter Walkemeyer*, Sedgman Onyx Pty Ltd., Perth, Australia
*Greg Wetterau*, CDM Smith, Rancho Cucamonga, Calif.
*Richard Whiting*, Water Sector Consulting LLC, Tampa, Fla.
*Zita L.T. Yu*, Jacobs, Los Angeles, Calif.

This page intentionally blank.

# Preface

As global populations continue to grow and traditional water supplies become increasingly stressed, many municipalities have been considering the ocean as a potential source of sustainable and long-term water supply. Continuing technological advancements in water treatment, water quality monitoring, energy recovery, and renewable energy supply have been transforming seawater desalination from a niche water supply for wealthy desert regions into a realistic and competitive alternative for many coastal areas facing supply challenges in their futures. In the United States, most desalination facilities have historically treated brackish groundwater, with an increasing number treating municipal wastewater for potable reuse; however, seawater desalination plants globally outnumber brackish water plants by nearly 3 to 1. Desalination facilities treating ocean water are currently operated in the United States in Florida, California, Massachusetts, and Washington, among other locations, with moderate- and large-scale facilities being planned or evaluated in coastal areas across the country.

With the growing interest in and increasing need for desalination as a source of water supply, this manual provides guidance for municipalities, water supply planners, and industrial water users considering ocean desalination as an alternative in their supply portfolios. The manual is an update to the first edition AWWA Manual M61 published over a decade ago and includes updated information on the rapidly evolving industry, as well as new chapters on construction and procurement, operations and maintenance, and case studies.

The manual provides a summary of the current state of practice for seawater desalination, including an overview of technologies, delivery methods, and key issues associated with the implementation of seawater desalination. Environmental protection has become a major focus for desalination facilities as plants become larger and more widespread and society looks for sustainable solutions to our long-term challenges. While there are inherent environmental benefits to the introduction of new freshwater supplies into water stressed areas, there have also been concerns raised over the environmental impacts of seawater intakes, concentrate discharge, and greenhouse gas emissions. This manual includes a detailed discussion of strategies being used or considered to address these concerns while maintaining the social and environmental benefits of new freshwater supplies and ensuring that these sustainable supplies remain truly sustainable for the foreseeable future.

Greg Wetterau
Subcommittee Chair, M61 Desalination of Seawater

This page intentionally blank.

Chapter 1

# Seawater Desalination Overview

*Second Edition Author: Greg Wetterau*
*First Edition Authors: Sandeep Sethi, Greg Wetterau*

## 1.1 INTRODUCTION

With global populations continuing to grow and freshwater supplies becoming increasingly stressed, many coastal areas have been turning to oceans and saline seas as a source of water supply. Three-quarters of the globe is covered with water; however, the vast majority is too saline for human consumption, and less than 0.3 percent is considered a renewable freshwater supply. More than half of the population in the United States lives within 50 mi (80 km) of an ocean gulf coast, and 40 percent of the world's population lives within 60 mi (100 km) of a type of saltwater coast, so the ability to convert seawater into a reliable source of drinking water supply holds tremendous interest, both within the United States and throughout the world. This type of adaptation is considered part of a growing blue economy (United Nations 2021) and can help boost sustainable development while relieving the pressures placed on overtaxed freshwater resources resulting from growing populations and increased urbanization.

The over-withdrawal of freshwater resources not only jeopardizes the communities that rely on them but degrades their surrounding habitats and disrupts essential ecosystem processes. As these finite resources become depleted, water utilities and their communities face the risk of elevated vulnerability in times of extreme drought or natural disasters unless alternative sources are secured. Seawater desalination presents an opportunity to provide a consistent and reliable water supply, reinforcing water security in coastal regions. As coastal populations continue to rise, with growth also predicted for the foreseeable future, seawater desalination provides a valuable tool to accommodate the unavoidable spikes in water demand.

During most of the 20th century, the high cost of desalination, including the high-power demand needed to operate the plants, made it less attractive than freshwater supplies, even where those supplies were hundreds of miles away. Large construction projects like the California State Water Project—which began in the 1960s and conveyed water to 27 million Californians using over 700 mi of pipelines, canals, tunnels, and reservoirs—focused on storing and conveying fresh water over long distances rather than converting local saline water bodies into sustainable freshwater supplies. Throughout much of the century, the use of seawater desalination remained limited to wealthy geographies with extreme water needs. With more recent advances in energy recovery, improvements in desalination membranes, and an increasing focus on sustainability, seawater desalination has become an increasingly viable alternative both in the United States and throughout the world.

Along with the improving cost of desalination technologies, the recent growth of renewable energy has been further reducing the cost of desalination, lessening the environmental footprint and allowing desalination to be implemented in a rapidly growing number of locations worldwide. Figure 1-1 shows that the worldwide desalination capacity increased more than 5-fold between 2000 and 2020. In the United States, most desalination facilities treat brackish water or are membrane softening plants; however, globally, the seawater desalination plants outnumber brackish water plants by nearly three to one (GWI 2024).

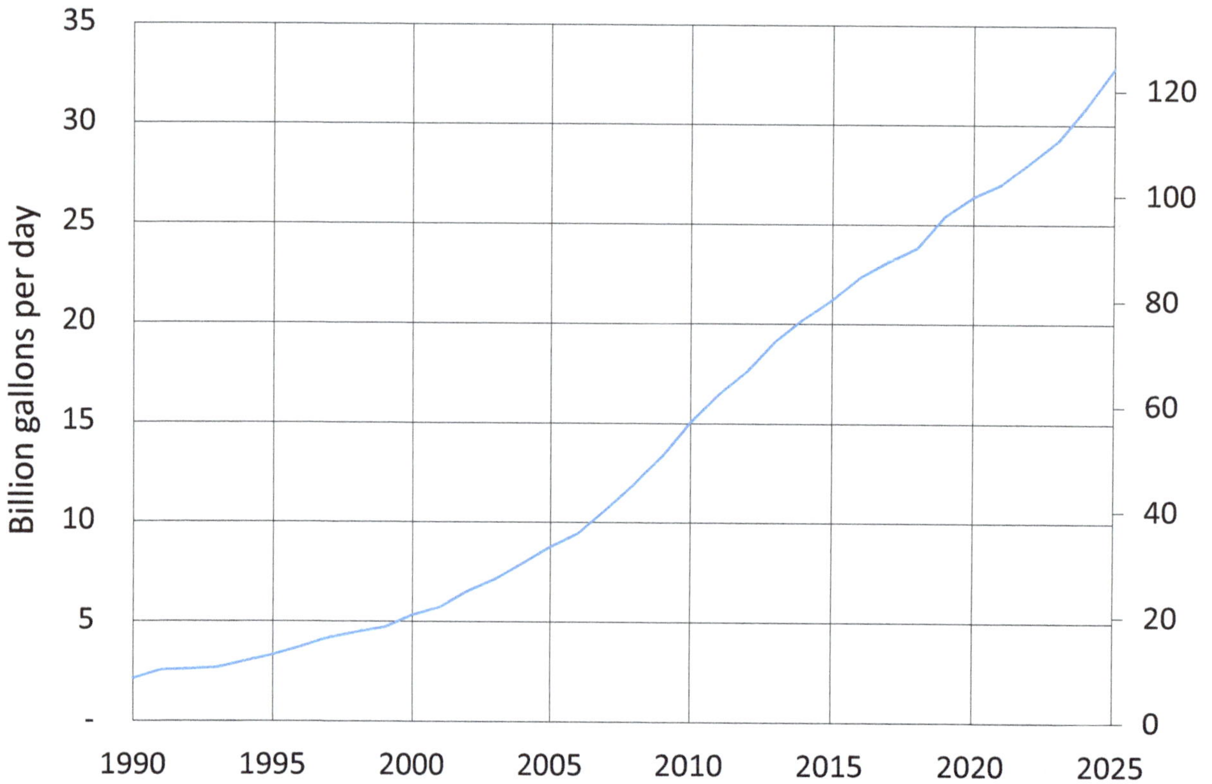


*Adapted from* GWI 2024

**Figure 1-1 Global growth of seawater desalination facilities**

The global use of desalination today exceeds 30 billion gal per day (bgd), employing core technologies developed primarily in the United States with more than \$2.4 billion (2024 US\$) of federal research funds between 1959 and 1974 (National Research Council 2008). While this production accounts for only 1 percent of all public water usage on the planet, desalination provides a critical supply alternative for water stressed areas, using an essentially limitless source of water. According to a 2019 report from the World Bank, there are over 20,000 desalination plants in more than 150 countries supplying over 300 million people with drinking water. With a few exceptions, new desalination plants have predominantly used seawater as their source of supply. Today, 61 percent of the global desalinated water production is from seawater, while only 21 percent uses brackish groundwater supplies. A breakdown of the source waters for existing desalination facilities is shown in Figure 1-2.

As of 2020, a total of 2.2 bgd (8.5 million $m^3/d$) of desalination capacity had been installed in the United States (GWI 2024). Data on desalination use in the United States are currently limited to municipal facilities, which make up the majority of installed capacity but not the total number of installed plants. The current municipal facilities in the United States all use membrane technologies and are overwhelmingly brackish water plants.

Table 1-1 lists the installed seawater desalination plants in the United States with capacities larger than 0.1 million gal per day (mgd). In addition to these plants, there are numerous smaller capacity plants operating in coastal communities, islands, or on offshore platforms that provide water for various industrial, municipal, and community needs. Notable small capacity plants include a 40,000-gpd facility in Monterey, Calif.; a 20,000-gpd facility on San Nicholas Island, Calif.; and 12 facilities installed in the San Juan Islands in Washington, with capacities up to 25,000 gpd (Lopez Island) (Mayo 2009). There are hundreds of operational desalination facilities located on offshore oil platforms off the coasts of Florida, Texas, Louisiana, and California. While the majority of desalination facilities in the United States are small capacity, large-scale plants are operational in both Tampa, Fla., and Carlsbad, Calif., with midcapacity plants operating in Santa Barbara,

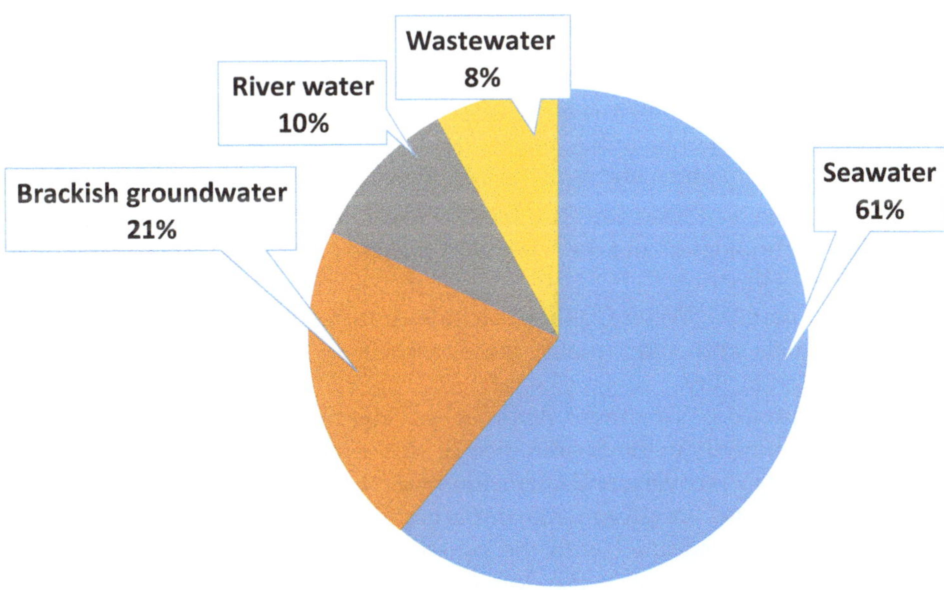


*Adapted from* International Desalination and Reuse Association 2024

**Figure 1-2 Total worldwide installed desalination capacity by feedwater category**

**Table 1-1 Operational seawater desalination facilities in the United States**

| Location | Capacity, mgd ($m^3$/d) | Supply Type | Finished Water Use |
|---|---|---|---|
| Avalon, Calif. | 0.23 (870) | Beach wells | Municipal |
| Carlsbad, Calif. | 50 (189,000) | Open ocean | Municipal |
| Santa Barbara, Calif. | 3 (11,000) | Open ocean | Municipal |
| Diablo Canyon, Calif. | 0.6 (2,300) | Open ocean | Industrial |
| Gaviota, Calif. | 0.4 (1,500) | Open ocean | Industrial |
| Morro Bay, Calif. | 0.6 (2,300) | Offshore wells | Municipal |
| Moss Landing, Calif. | 0.5 (1,900) | Tidal slough | Industrial |
| Sand City, Calif. | 0.3 (1,100) | Beach wells | Municipal |
| Tampa, Fla. | 25 (95,000) | Open ocean | Municipal |
| Stock Island, Fla. | 4 (15,000) | Subocean wells | Municipal |
| Marathon Key, Fla. | 4 (15,000) | Subocean wells | Municipal |
| Swansea, Mass. | 2 (7,600) | Tidal impacted river | Municipal |
| Brockton, Mass. | 5 (19,000) | Tidal impacted river | Municipal |

Calif.; Stock Island, Fla.; Marathon Key, Fla.; Swansea, Mass.; and Brockton, Mass. Other large-scale projects are under development in Monterey and Dana Point, Calif., as well as Brownsville and Corpus Christi, Tex. A number of desalination plants in California have also been delayed or cancelled due to perceived negative impacts on the environment or to rising costs associated with addressing these environmental concerns. As coastal municipalities in the United States continue to consider implementing larger seawater facilities, it will be essential to ensure that they are constructed and operated in an efficient and reliable fashion without adversely impacting fragile coastal environments. Large-capacity, highly efficient seawater desalination facilities have been successfully implemented within the last 2 decades in Australia, Singapore, South American countries, and several countries in the Middle East. In the United States, there are currently over two dozen new seawater projects in various stages of development.

The purpose of this manual of practice is to identify lessons learned from recent studies and seawater desalination projects around the world and to use these to provide guidance for seawater desalination facilities that are reliable, economical, and environmentally sound. The manual includes major updates from an earlier 2010 manual, including three entirely new chapters, and is organized into the following chapters:

**Chapter 1**—"Seawater Desalination Overview" includes information on desalination technologies and key issues associated with the implementation of seawater desalination.

**Chapter 2**—"Water Quality" includes information on seawater quality, water quality goals, and water quality issues associated with operation of a seawater desalination facility.

**Chapter 3**—"Treatment Approaches" discusses treatment technologies used most commonly in the desalination of seawater, including pretreatment, posttreatment, energy recovery, and corrosion mitigation.

**Chapter 4**—"Environmental Impacts and Mitigation" discusses the key environmental issues associated with the construction and operation of seawater desalination facilities, as well as the state-of-the-industry in mitigation measures being used to address these concerns.

**Chapter 5**—"Cost of Treatment" provides a high-level discussion of seawater desalination costs, including both construction and long-term operation and maintenance costs.

**Chapter 6**—"Construction and Procurement of Desalination Facilities" discusses the most commonly used construction, contracting, and financing models used in the implementation of desalination plants.

**Chapter 7**—"Plant Operations and Maintenance" discusses key issues associated with the operation, maintenance, staffing, and optimization of desalination plants.

**Chapter 8**—"Case Studies" includes case studies for nine desalination facilities either in the United States or in other countries with designs that are relevant to implementation of desalination within the United States.

## 1.2 DESALINATION TECHNOLOGIES OVERVIEW

Desalination processes can be divided into two broad categories: membrane separation and thermal evaporation. Membrane-based desalination processes typically employ mechanical pressure, electrical potential, or a concentration gradient as the driving force across a semipermeable membrane barrier to achieve physical separation. Thermal desalination processes employ heat to evaporate the water from a salt solution, and the water vapor is then condensed and recovered.

Thermal technologies were the only options available for seawater desalination until reverse osmosis (RO) membranes were developed in Los Angeles in the early 1960s. Since then, RO membrane processes have steadily been improved, and their efficiency has increased to the point that they are now the technology of choice for most seawater desalination applications, including projects in the Middle East, where over half of the world's desalination plants are located.

Besides the established desalination technologies, there are several newer technologies that are nearing commercialization or undergoing active research and development. A discussion of the established membrane and thermal technologies is presented first, followed by a brief discussion of developing technologies. The remaining chapters in this manual focus on pressure-driven membrane applications, as this presently has the most applicability to seawater desalination in the United States.

## 1.3 MEMBRANE SEPARATION

Membrane desalination technologies have been designed around the ability of semipermeable membranes to selectively permit or minimize the passage of certain ions. Three fundamental driving forces can be used in membrane desalination systems, including pressure, electric potential, and concentration gradient. RO and nanofiltration (NF) are pressure-driven processes. Electrodialysis (ED) and electrodialysis reversal (EDR) are electric potential–driven processes. Forward osmosis (FO) is a concentration-driven process. Membrane distillation (MD) is a hybrid process using both evaporative and membrane technologies, but it is ultimately pressure differences that drive the separation in MD. Each of these processes is summarized in Table 1-2, with more information provided in the subsequent sections of this chapter.

While all of these membrane-based technologies can be used for desalination, RO is the primary membrane technology used when treating seawater. NF and ED/EDR are also mature technologies; however, ED/EDR are typically not cost competitive for desalination of seawater (Amjad 1996), and NF salt rejection efficiency makes it challenging to apply in seawater desalination for potable water production. NF has been adopted for sulfate removal in offshore oil production facilities, where low salinity water is not required. A novel approach employing a two-pass (NF) configuration was developed and tested for seawater desalination by the Long Beach Water Department in California; however, no full-scale projects have moved forward using this configuration. FO is a developing technology

**Table 1-2 Membrane-based desalination processes**

| Process | Driving Force | Usage | Section |
|---|---|---|---|
| Reverse osmosis | Pressure | Common for seawater desalination | 1.3.1 |
| Nanofiltration | Pressure | Common for low-to-moderate salinity water | 1.3.2 |
| Electrodialysis and electrodialysis reversal | Electric potential | Moderate usage for low-to-moderate salinity water | 1.3.3 |
| Forward osmosis | Concentration gradient | Limited | 1.5.1 |
| Membrane distillation | Vapor pressure | Limited | 1.5.2 |

used primarily in the food and beverage industry, and it has not been commercialized for large-scale or for seawater applications. Similarly, MD is a developing technology, with limited use in full-scale applications.

### 1.3.1 RO

**Process overview.** Desalination through RO is a well-established and nonproprietary unit process that currently represents the state-of-the-art of desalination technology for a number of reasons. In addition to the ability to reject a variety of contaminants, the RO treatment process generally has lower energy consumption, lower feed water flows, and no thermal impacts in the concentrate discharge in comparison to thermal desalination processes. In contrast, water produced using thermal processes must be cooled before it can be distributed for human consumption. Improvements in membranes and energy recovery devices (ERDs) used for seawater RO (SWRO) have improved the overall process efficiency, thereby lowering the costs associated with treatment. RO is based on overcoming the natural phenomenon of osmotic pressure, which occurs when a semipermeable membrane separates two solutions with different concentrations of ions. In the natural osmosis process, the osmotic pressure created by the difference in concentration drives the flow of water from the dilute solution to the concentrated solution, diluting the water on the higher concentration side of the membrane until chemical equilibrium is reached. In RO, the flow of water is reversed, increasing the amount of water on the low salinity side of the membrane, with the application of an external hydraulic force (pressure) that is greater than the natural osmotic pressure. As the water on the high concentration side of the membrane becomes further concentrated, osmotic pressure will increase and greater pressure must be applied to continue moving water to the lower salinity side (see Figure 1-3).

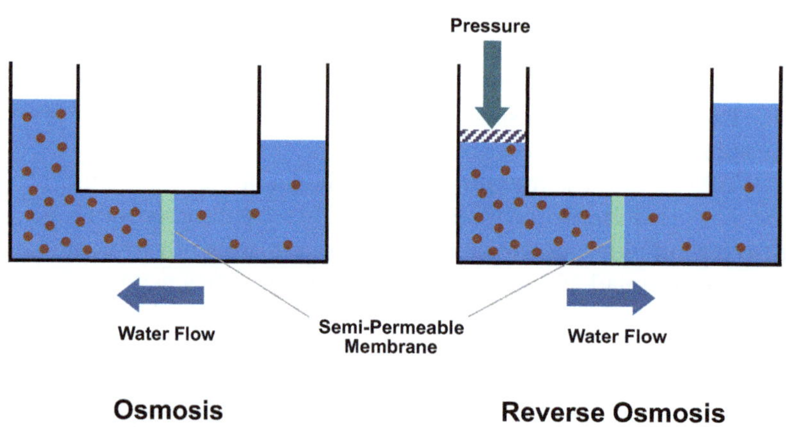


**Figure 1-3 Basic concept of osmosis and reverse osmosis**

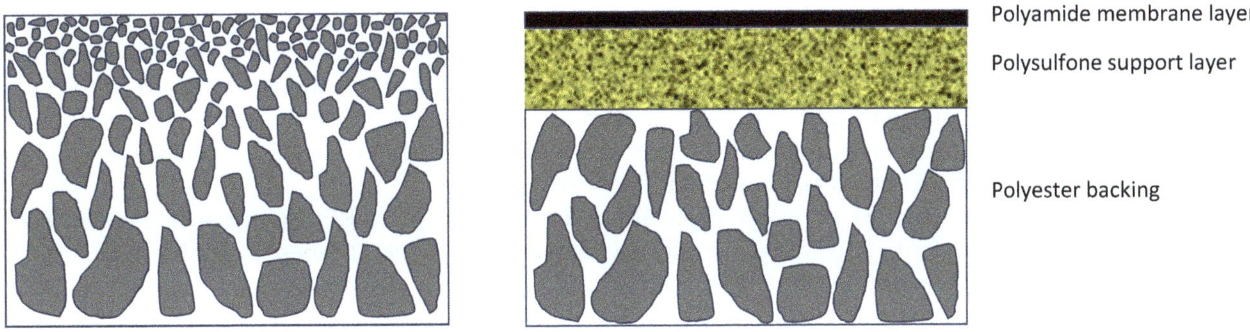


**Figure 1-4 Structure of cellulose acetate (CA, left) and thin-film composite (TFC, right) membranes**

**Membranes.** RO membranes are designed to retain salts and low-molecular-weight solutes while allowing water to pass through. The original asymmetric cellulose acetate (CA) membranes, developed in the 1960s, were less permeable than modern thin-film composite (TFC) membranes and required higher driving pressure, in excess of 1,200 psi or 8.3 megapascals (MPa) for seawater at typical plant production rates. Additionally, the ability of CA membranes to reject salts was originally less than current materials.

CA membranes utilized an asymmetric structure, while the TFC contained multiple layers made from different materials. These structures are illustrated in Figure 1-4. In the asymmetric configuration, the membrane consists of the same material throughout with a dense layer at the top and porous layer beneath. In contrast, the TFC membrane consists of a thin but dense layer of one material over a porous support consisting of a different material.

Today, there are a variety of modified and improved blends of CA membranes available to the desalination industry, but these membranes are rarely used in large-scale desalination applications. CA membranes can tolerate continuous exposure to low concentrations of chlorine (0.1–0.5 mg/L at 25°C), which is an advantage for biofouling control in seawater applications. They are, however, susceptible to hydrolysis, which compromises the membrane's salt rejection performance. Hydrolysis of CA membranes is accelerated if the operating pH is less than approximately 4 or greater than approximately 7 and temperatures greater than 30°C (Mallevialle et al. 1996). Therefore, pH depression into this range is needed for seawater desalination with CA membranes.

The development of TFC membranes provided greater salt rejection and higher water production per unit membrane area. TFC membranes are made by combining a thin, dense membrane film with a porous underlying material that provides structural support. The thin film typically consists of aromatic polyamide, and the bottom support layer is typically polysulfone. Most of the solute rejection occurs at the thin nonporous film, and its small thickness can significantly reduce the pressure required to drive water through it in comparison to CA membranes. TFC membranes are stable over a broad pH range (2–11) and can withstand temperatures as high as 45°C (113°F). However, unlike the CA membranes, they are susceptible to degradation by strong oxidants such as free chlorine. Although the degradation rate caused by free chlorine is a function of pH, membrane materials generally deteriorate upon exposure (sometimes catastrophically).

High pressures are required to overcome the osmotic pressure of the salts and minerals, the resistance from the membrane material, and other associated system losses. SWRO membranes are typically operated at feed pressures of approximately 800–1,000 psi (5.5–6.9 MPa); however, typical SWRO elements can tolerate pressures up to 1,200 psi (8.3 MPa), and ultra high–pressure elements are available that can be operated at pressures up to 1,800 psi (12.4 MPa), allowing for higher recoveries or fluxes to be achieved if needed. RO membranes are capable of rejecting contaminants as small as 0.1 nm; however, the process of water transfer is mostly diffusion controlled rather than convection controlled, as is the case with MF and UF. In addition to the effects of the major ion matrices, mass transfer of ions through RO membranes is also impacted by broader water quality characteristics, such as temperature and pH.

The amount of water typically recovered using SWRO membranes ranges from 35–60 percent, and commercially available SWRO membranes typically reject 99.5–99.8 percent of the total dissolved solids (TDS) in the feed water. Removal of a few constituents, such as boron and chloride, is sometimes not as great as might be required at every facility (see Chapter 2). If product water goals are not met, additional treatment may consist of two-pass RO, in which a portion or all of the product water (permeate) produced in the first pass is treated again in a second pass. New membranes with improved boron rejection are being developed by SWRO manufacturers to avoid the need for two-pass treatment; however, other water quality goals besides boron rejection may also impact the need for a two-pass system. An optimized SWRO design will therefore depend on the feed water quality, system operating conditions, and specific finished water quality requirements.

Because membrane processes are based on physical separation, they do not require thermal energy to vaporize the water (with the exception being MD, discussed later in this chapter). As a result, the energy consumption for treatment components of an SWRO plant typically falls in a range of 10–20 kilowatt-h (kW•h) per 1,000 gal (2.6–5.3 kW•h/m$^3$). In comparison, total energy used for thermal desalination treatment processes can range from 20 to 60 kW•h/1,000 gal (5.3–16 kW•h/m$^3$), depending on the unit processes (GWI 2024; Eke et al. 2020; Wade 2001).

ERDs have become standard in SWRO applications. These devices can recover from 25 to over 45 percent of input energy for SWRO. Examples of such devices as presented in Chapter 3 include Pelton wheels, work exchangers, pressure exchangers, and hydraulic turbo-exchangers.

One of the greatest challenges for membrane desalination processes is fouling and scaling of the membranes. Fouling can occur as a result of inadequate pretreatment or measures for reduction of particulate, colloidal, or organic matter to tolerable levels or biological growth in the membrane pressure vessels. Scaling results from precipitation of sparingly soluble salts in the system like silica or calcium carbonate and tends to be less of a concern in seawater desalination than in brackish water systems that run at higher recoveries. Compounds such as calcium carbonate and calcium sulfate in the feed water may, in some cases, contribute to limiting the recovery of the SWRO process. Acid or scale inhibitors (also known as antiscalants) may be added to reduce alkalinity and prevent formation of scale, allowing for higher recovery than otherwise possible. Due to high levels of particulates and the generally aerobic state of seawater, SWRO plants require comprehensive pretreatment and chemical conditioning of the feedwater for successful operation. This is particularly true when open ocean intakes are used to supply the desalination plant, but less pretreatment may be required where subsurface intakes are employed. Intake approaches are discussed in more detail in Chapters 3 and 4.

### 1.3.2 NF

NF is typically used to soften water and remove disinfection byproduct (DBP) precursors such as dissolved organic matter; NF is typically not used for seawater desalination, although unique configurations of two-pass NF have been successfully used to desalinate seawater.

Nanofiltration uses semipermeable membranes and a driving force of hydraulic pressure; however, in comparisons to RO, NF membranes typically have a higher molecular weight cut-off (MWCO). NF membranes remove a high percentage (90–98%) of divalent ions (i.e., those associated with hardness), but removal of monovalent ions is somewhat limited (typically 60–85%).

Because a higher concentration of monovalent ions can pass through the NF membrane, the osmotic pressure is lower compared to RO. This, combined with a more permeable membrane skin layer, reduces the hydraulic pressure requirements to 500–700 psi (3.4–4.8 MPa) for seawater applications. Recognizing these advantages, the Long Beach Water Department in California developed and patented an innovative two-pass nanofiltration method for the desalination of seawater. This is discussed further in Chapter 3.

### 1.3.3 Electrodialysis (ED)/Electrodialysis Reversal (EDR)

ED and EDR processes use ion-selective membranes and an electrical potential as a driving force to separate charged species from water. Pressure-driven systems (RO and NF) selectively pass water through a membrane and retain dissolved salts in the concentrate. In contrast, ED and EDR use an electrical potential to draw dissolved ions through a set of membranes (cations to one side, anions to the other), while the deionized water passes between the membranes and is ultimately recovered.

An ED stack consisting of alternating layers of cationic and anionic ion-selective flat sheet membranes creates channels of desalted product water and concentrated reject water. Cations migrate to the cathode and anions migrate to the anode, while cation-selective membranes allow only cations to pass and anion-selective membranes allow only anions to pass. The net effect is to remove the salt from every other cell.

EDR, a modification of the ED process, periodically reverses the polarity of the applied electrical potential on the stack to minimize the effects of inorganic scaling and fouling. By reversing the direction of the ion migration, the product channels and concentrate channels are switched and imbedded or impacted ions are removed from the surface of the membranes. This allows the EDR system to operate at higher recoveries compared to ED.

ED/EDR processes are typically not used for seawater desalination because with higher salinities, the ED/EDR process generally becomes less efficient than other membrane-based desalination technologies. Additionally, bacteria, nonionic constituents, and residual turbidity are not affected by this process and therefore remain in the product water, requiring additional treatment before drinking water standards are met. Since ED/EDR processes are not typically used for seawater desalination, they will not be discussed further in this manual; however, more information about them can be found in AWWA Manual M38 *Electrodialysis and Electrodialysis Reversal*.

## 1.4 THERMAL EVAPORATION

Thermal desalination technologies work by evaporating water from a saline solution and then condensing the vapor (steam) to produce distilled water. All large-scale thermal processes involve heating water to its boiling temperature to produce the maximum amount of water vapor. Following the ideal gas law, the pressure of the system is typically decreased so that the temperature required for boiling is reduced. Commercially available distillation systems are designed to allow for "multiple boilings" in a series of vessels that operate at successively lower temperatures and pressures.

Thermal technologies that are used for desalination include multistage flash (MSF), multiple effect distillation (MED), and vapor compression (VC). MSF and MED systems typically use direct heat exchange from steam as the energy source for evaporation, while VC systems use the heat from the compression of the vapor as the energy source for evaporation. Thermal processes can produce water with very low salt concentrations (TDS levels of 10 mg/L or less) from TDS levels as high as 60,000–70,000 mg/L TDS; however, there are limitations associated with distillation processes for seawater desalination.

One of the most significant limitations of thermal technologies is the energy requirement of the vaporization step. High levels of salts result in boiling point elevation, and the energy required to vaporize seawater typically ranges from around 20–100 kW•h/1,000 gal (5–26 kW•h/m$^3$) of fresh water produced (Wade 2001). It should be noted that these thermal energy requirements are in addition to the electrical energy required for the other aspects of the process. Often, large distillation plants are coupled with steam or gas turbine power plants, making use of low-grade heat to reduce power input requirements. Thermal technologies are more commonly used in the Middle East, where energy costs are relatively low, the large land requirements are not cost prohibitive, and ecological permitting requirements are less stringent. There has long been interest in using solar energy as a source of heat for accomplishing the evaporation in distillation, but suitable technologies for a large-scale project are not yet available as of this publication.

Operational issues for thermal desalination include corrosion and scaling. Because seawater is highly corrosive in nature, special alloys like cupronickel alloys, aluminum, and titanium are used most commonly in desalination with distillation processes. These alloys contribute significantly to the capital cost of a distillation plant, particularly with the large surface area required for efficient distillation. The scaling of sparingly soluble salts at elevated temperatures on the inner walls of pipes and equipment is another operational issue that reduces the heat transfer efficiency of the heat exchangers and increases the overall energy required for distillation. Also, additional permitting concerns may arise because concentrate discharged from a thermal distillation process has a higher temperature than the ambient water in the discharge location. While the cost of thermal desalination is often considerably higher than RO, very little pretreatment is required ahead of thermal processes, and the product water quality is extremely high (less than 10 mg/L TDS), which eliminates the need for additional treatment to address boron, chloride, or bromide concerns.

### 1.4.1 MSF

MSF accounts for the greatest installed thermal distillation capacity worldwide. In the MSF process, water is heated in a series of stages, each with successively lower pressures and temperatures. Typically, MSF plants can contain from 15 to 25 stages. Vapor generation or boiling caused by reduction in pressure is known as flashing (illustrated in Figure 1-5). As the water enters each stage through a pressure-reducing nozzle, a portion of the water is flashed to form vapor. In turn, the flashed water condenses on the outside of the condenser tubes and is collected in trays. As the vapor condenses, the latent heat is used to preheat the seawater that is being returned to the main heater where it will receive additional heat before being introduced to the first flashing stage. The condensate collected in each stage forms the product, and the whole process is driven by a subatmospheric pressure gradient through the stages.

Evaporation or flashing of a small portion of the feed continues in each successive stage at a lower pressure. The MSF process generates and condenses vapor in the same stage or effect. The range of recoveries for conventional MSF desalination processes is limited to about 10–30 percent for seawater desalination.

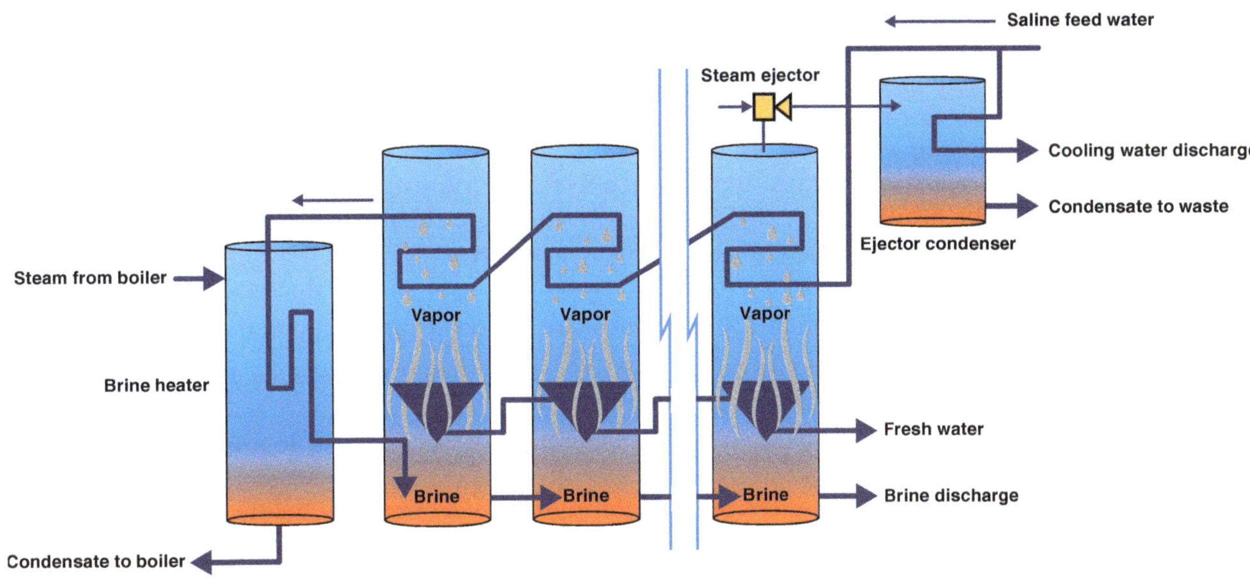


*Source:* Greg Wetterau

**Figure 1-5 Multistage flash distillation**

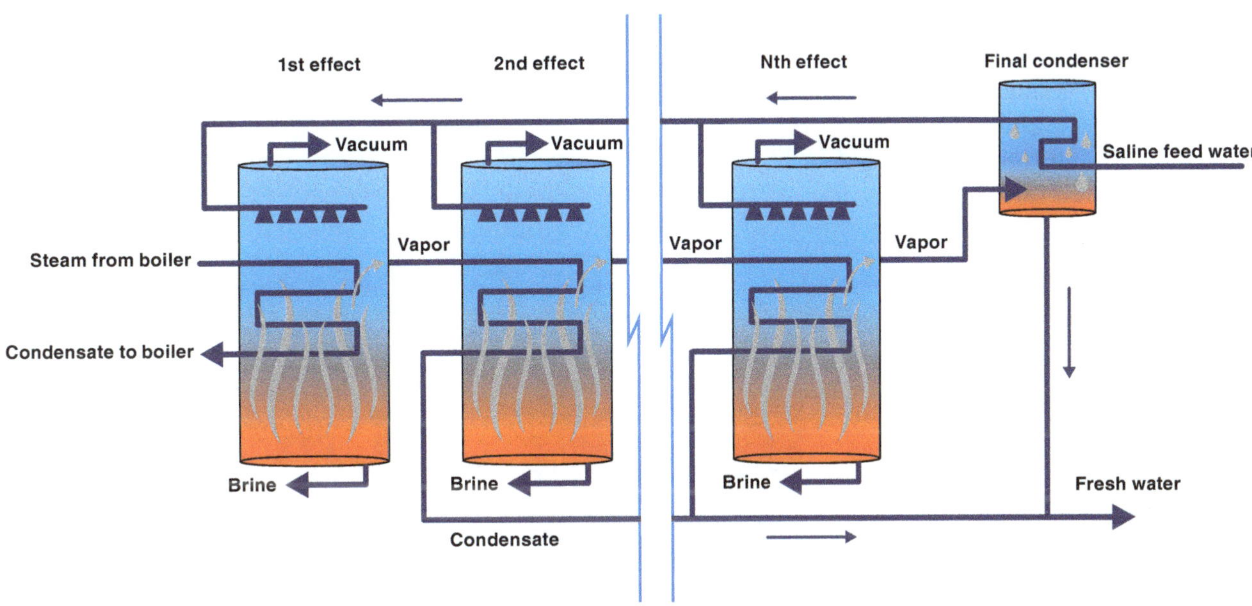


*Source:* Greg Wetterau

**Figure 1-6 Multiple effect distillation**

## 1.4.2 MED

The MED process, like the MSF process, uses multiple vessels (or effects) arranged in series with reduced ambient pressure in each subsequent effect. Typically, 8 to 16 effects are used in MED to minimize the energy consumption. The feed water is distributed on the outside of the evaporator tubes in a thin film (see illustration in Figure 1-6) to promote rapid boiling and evaporation. Steam is condensed on the colder inside surface. Vapor produced by evaporation is condensed in a way that uses the heat of vaporization to heat the remaining saline solution at a lower temperature and pressure in each succeeding effect, allowing water to undergo multiple boilings without supplying any additional heat after the first effect. Thus, the vapor produced in each effect is used to heat the feed water in the next effect. This not only reduces the energy required for distillation but also the overall electrical power consumption. As a result, energy costs for operating an MED plant can be half of what is required for an MSF plant.

The steam generated in the final effect is typically at a pressure and temperature too low to be of further use. MED systems normally condense this steam using an external cooling source to remove the heat of condensation.

Energy is required in an MED system

- to create steam of sufficient pressure to drive evaporation in the first stage;
- to power vacuum systems to reduce the boiling pressure in the downstream effects (if operated at low temperatures); and
- to pump influent water through the heat exchangers to the evaporator(s), to recirculate the concentrate within each evaporator stage, and to pump the condensate and concentrate through the heat recovery before exiting the system and then cooling water to condense the steam from the final stage.

Energy efficiencies may be gained via the combination of the evaporator systems with available low-pressure or waste steam/heat sources or by the addition of efficiency enhancement devices to the conventional MED system. The range of recoveries for conventional MED is limited to approximately 20–35 percent for seawater desalination.

## 1.4.3 VC

Heat for evaporation in VC systems is provided by one of two approaches: mechanical vapor compression (MVC) or thermal vapor compression (TVC). An illustration of the former is provided in Figure 1-7. MVC systems use electricity while TVC systems use

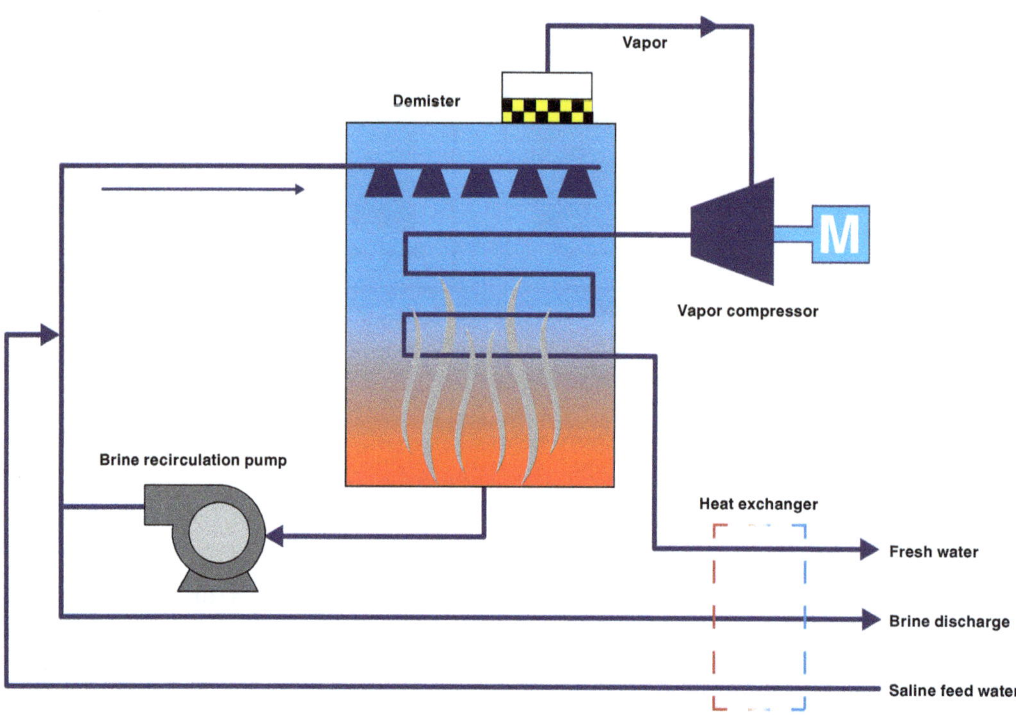


*Source:* Greg Wetterau

**Figure 1-7 Vapor compression**

high-pressure steam to compress the water vapor created from distillation to higher pressure and temperature so that it can be returned to the evaporator and used as a heat source. The vapor compression process is well established and is used for seawater desalination as well as treating RO concentrate for residuals management. Vapor compression systems typically have recoveries in the range of 40–50 percent for seawater desalination. VC desalination systems require more energy than MED or MSF systems, with specific energy usage 50–100 percent higher than MSF.

## 1.5 NOVEL DESALINATION PROCESSES IN DEVELOPMENT

### 1.5.1 FO

As in the case of RO and NF, FO employs a semipermeable membrane to separate water from a saline solution; however, instead of using external hydraulic pressure to create the driving force for water transport through the membrane, the FO process employs a natural pressure gradient provided by a higher salinity draw solution (such as ammonium carbonate or specially prepared magnetic nanoparticles). The higher osmotic pressure of the "draw" solution causes water to move towards it through a membrane. Fresh water is then separated from the draw solution using an additional separation process that can vary depending on the nature of the draw solute. The separated draw solutes are either recovered and reused in the FO process or discharged.

Osmotic driving forces in FO can be significantly greater than hydraulic driving forces used in RO. Use of a suitable draw solution with very high osmotic pressure-driving forces can be used in principle to generate high water fluxes and recoveries. The FO process, once fully developed and commercialized, is expected to have potential advantages due to its relatively low fouling potential, low energy consumption, and simplicity. Identification of appropriate draw solutions and development of efficient membranes are two of the most pressing challenges for FO. An effective draw solute should have the following characteristics:

- high osmotic efficiency, meaning that it has to be highly soluble in water and have a low molecular weight in order to generate a high osmotic pressure;
- nontoxicity, as trace amounts may be present in the product water;
- chemical compatibility of the membrane; and
- easy and economical separation from recovered water.

Commercial RO membranes are not suited for the FO process because of relatively low product water fluxes. This low water flux is due mainly to internal concentration polarization within the porous support layer of the membrane, which alters the effective driving force across the active layer of the membrane and thereby limits water flux. One of the important tasks for future research is the development of a semipermeable FO membrane having high salt rejection and minimal internal concentration polarization to realize higher product water fluxes (see Figure 1-8).

FO has been successfully adapted to industrial treatment in the food and beverage industry; however, its use in treatment of high salinity source waters remains at a precommercial level. Early research included bench-scale FO units built and operated at a Yale University laboratory, New Haven, Conn. (McCutcheon and Elimelech 2006), supported by Office of Naval Research (Award No. N000140311004). The draw solution employed in these studies consists of highly concentrated ammonium carbonate prepared by mixing ammonia and carbon dioxide gases. Upon heating to approximately 60°C, ammonium carbonate decomposes back into ammonia and carbon dioxide gases, leaving behind the desalinated water. This separation needs to be essentially complete due to

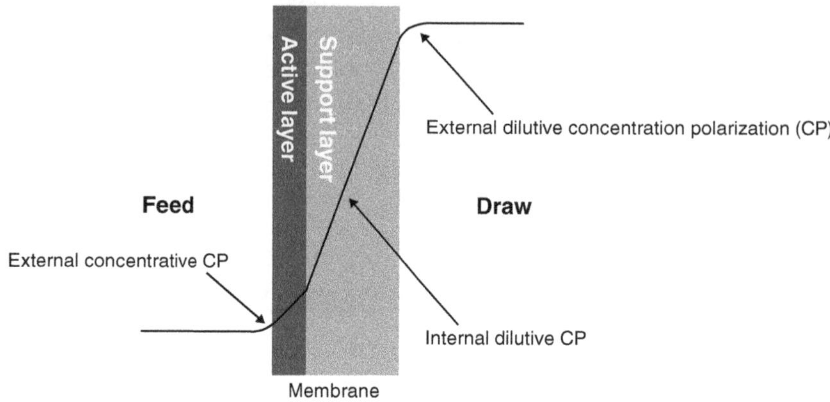


**Figure 1-8 Salinity concentration profile during forward osmosis (FO) process**

limits on ammonia in drinking water. Thermal recovery of ammonia and carbon dioxide from the draw solution requires energy, so the process may be suited for applications where low-grade heat is available.

FO has also been studied for use as pretreatment for RO in several novel applications including water reclamation and nutrient recovery (Holloway et al. 2007; Cath et al. 2005). Several hybrid processes have been developed, including osmotic dilution before desalination, which reportedly can significantly reduce the energy demand of desalination in some applications (Cath et al. 2009). A 500 $m^3$/d (132,000 gpd) desalination facility was started up in 2024 in Kona, Hawaii, using FO with solar-thermal collectors to demonstrate a novel green desalination technology at the Natural Energy Laboratory of Hawaii. Finally, FO has been proposed as a method of energy generation in blending highly concentrated desalination brine with fresh wastewater flows before discharging into the ocean. Such an approach would employ two waste products to produce electrical power and reduce the overall energy footprint of the desalination facility. A demonstration facility using this technology has been operating at the Fukuoka Desalination Plant in Japan since 2010 (Makabe et al. 2021).

## 1.5.2 MD

MD is a hybrid process using principles of both membrane separation and thermal distillation. MD involves evaporation of water from a saline solution and transport of the water vapor through the pores of a hydrophobic membrane. The membrane allows water vapor to pass through but prevents the solution from passing through. Water vapor is transported across the membrane in response to a change in partial pressure across the membrane due to a thermal gradient. The clean vapor is subsequently carried away from the membrane and condensed as pure water either within the membrane package or in a separate condenser system.

MD differs from pressure-driven membrane technologies in that, rather than applying pressure to force liquid through a membrane, the driving force for desalination is the difference in vapor pressure of the liquid across the membrane. Increasing the temperature of the liquid increases the vapor pressure and results in increased membrane penetration rates. The efficiency of an MD process largely depends on the feed water quality, system design, and heat recovery from the permeate stream. MD has been reported to run at relatively low temperatures (approximately 70°C or 158°F) and thus can utilize waste heat or low-grade heat sources. The energy source for feed water heating and/or for a vacuum system to sweep away the vapor may be low-grade thermal energy like that supplied by low-pressure steam, waste heat, solar energy, or geothermal energy.

Potential advantages of MD are the ability to use low-grade heat, minimal pretreatment needs, and negligible scaling or precipitation concerns. Challenges include need of waste heat for economic feasibility, membrane fouling, and membrane degradation due to loss of hydrophobicity. MD technology is currently in the development and demonstration phase.

A variety of arrangements and configurations can be used to induce the vapor through the membrane and recover condensate and product water. Common to all concepts is that the feed water directly contacts the membrane. Condensation can be achieved using one of four process configurations:

- Air-gap membrane distillation: This configuration, which is the most common and most versatile arrangement, provides an air gap after the membrane, followed by a cool surface for condensation to occur.
- Direct-contact membrane distillation: The cool condensing solution (pure water) directly contacts the membrane and condenses the vapor as it passes through the membrane, where the coolant liquid typically flows counter-currently to the feed water.
- Sweep-gas membrane distillation: A sweep gas pulls the water vapor out of the membrane gap for subsequent condensation outside of the membrane package.
- Vacuum membrane distillation: A vacuum is applied to the membrane space to pull the water vapor out of the system.

### 1.5.3 Freeze/Thaw

The freeze/thaw approach to desalination is similar to thermal desalination in the fundamental concept of relying on a phase change to achieve separation. In the case of freeze/thaw, the phase change is from liquid to solid. Ice crystals exclude salt from their structure, and the salt can then be separated as a brine from the ice. A key aspect of the process is that the energy required for the phase change from water to ice is less than one-seventh the energy that is required for the phase change from water to vapor (however, practical thermal desalination processes like MSF and MED use much less energy than the heat of evaporation due to the use of multiple effects or stages, as discussed previously). Challenges with freeze desalination include implementing the proper washing and separation of the crystals without premature melting and/or recontamination with the excluded salt. Different configurations of freeze/thaw desalination systems that have been developed include direct freezing, indirect freezing, and absorption (AWWA 2004).

### 1.5.4 Capacitive Deionization (CDI)

Carbon aerogel is an ideal electrode material because of its high electrical conductivity, high specific surface area, and controllable pore size distribution (Ying et al. 2002; Yang et al. 2001). The Lawrence Livermore National Laboratory (LLNL) in California began its research into carbon aerogels and capacitive deionization technology (CDT) in the late 1980s. LLNL developed and optimized carbon aerogel materials, which multiplied the effective surface area of the deionization electrodes by a factor of 60,000, dramatically improving their capacity to attract and hold charged water constituents.

In the process of CDI, the high salinity solution flows between the electrode pairs and ions are adsorbed onto the surface of the porous electrodes by applying a low voltage electric field, which produces deionized water. The major mechanisms related to the removal of charged constituents during CDI are physisorption, chemisorption, electrodeposition, and/or electrophoresis. Unlike ion exchange, no additional chemicals are required for regeneration of the electrosorbent in this system. Adsorbed ions are desorbed from the surface of the electrodes by eliminating the electric field, resulting in the regeneration of the electrodes. The efficiency of CDI strongly depends upon the surface area and adsorption properties of electrodes.

CDI systems exhibit several advantages; they have a simple, modular plate and frame construction and a low energy requirement. However, current challenges include the limited adsorption capacity of carbon aerogel electrodes, slow kinetics of transport of ions into and out of the highly porous electrodes, relatively high costs of the CDI modules, and the fouling potential of the aerogel surface due to natural organic matter. CDI is still in the development stage with ongoing bench and pilot tests.

### 1.5.5 Supercritical Desalination (SCD)

Research by the Wetsus Center of Excellence for Sustainable Water Technology in the Netherlands (Ingo et al. 2009) has looked at the use of supercritical conditions in water to promote desalination. Supercritical conditions are achieved at elevated temperatures and pressures where the liquid and gas forms of water both exist and are indistinguishable from each other. Supercritical water is an extremely poor solvent of inorganic salts and allows dissolved salts to be removed through precipitation, which produces a solid or near solid waste stream and a purified product. The purity of the product will depend on the temperature and pressure of the supercritical fluid. SCD has been proposed as a pretreatment step for RO in seawater desalination, where it may be used to reduce the seawater salinity by 90 percent, thereby producing a water with TDS levels close to 3,500 mg/L using a pressure of 22 MPa (3,200 psi) and temperature of approximately 350°C (662°F). Low-pressure brackish RO membranes can then be used for further desalination with the theoretical energy requirements being comparable with conventional SWRO but with higher recoveries on the order of 70–80 percent. SCD technologies are still in development and have not been tested outside of the laboratory; however, these technologies may play a role in future desalination applications as they offer unique opportunities not currently available with either membrane- or evaporation-based technologies.

### 1.5.6 Halophytic Algae Desalination

The Lawrence Berkeley National Laboratory and the US Environmental Protection Agency National Risk Management Research Laboratory have been conducting research in the use of halophytic algae as a desalination tool (Sahle-Demessie et al. 2019). The approach uses salt-tolerant algae grown in a photobioreactor to concentrate salt within the algal cells. The algae can then be physically separated from the desalinated water using low-cost conventional filtration technologies. The concentrated algae produced in this process has been proposed as a biofuel source, helping to power the desalination process and offset further energy inputs into the treatment process. Researchers have been successful at removing up to 30 percent of the salt from prepared saline solutions using the halophytic algae. While this level of salt reduction is insufficient for producing drinking water, it could serve as a potential first step, significantly reducing the energy required to further desalinate the water to drinking water standards. This technology is still in the early stages of research development, and it has not been commercialized or adapted for large-scale municipal usage.

### 1.5.7 Offshore Submerged Desalination

A new approach to seawater desalination that has been gaining interest as a green technology is the use of offshore submerged systems employing hydraulic pressure from the deep ocean for desalination rather than direct pumping of feed water. While this approach employs conventional RO membranes to desalinate the water, the treatment configuration is different enough from a conventional desalination process to warrant a separate discussion. There are multiple companies currently marketing commercial systems, with demonstration tests and full-scale projects being developed in the United States and Middle East. The concept employs submerged pods or treatment modules tethered to the ocean floor at depths greater than 1,500 ft (450 m), where marine life abundance is low and environmental impacts are expected to be reduced. While the desalination process is driven by hydraulic pressure from the ocean, pumping is needed to convey desalinated water to the shore and to return concentrated brine back to the ocean.

Seawater at these low depths will be extremely cold (typically 4°C or 39°F), limiting the flux that can be achieved across the RO membranes to roughly 3–4 gal per $ft^2$ per day (5–7 $L/m^2$ h) using available head from the water column. These lower fluxes could decrease the quality of the product water; however, low recoveries (10–20 percent) could also be used to maintain similar product water quality to a conventional desalination plant. Technology providers have claimed the submerged desalination systems will use 40 percent less energy than a conventional desalination system while drastically reducing the environmental impacts.

## REFERENCES


Amjad, Z. 1996. *Reverse Osmosis: Membrane Technology. Water, Chemistry, and Industrial Applications*. New York: Van Nostrand Reinhold.

AWWA (American Water Works Association). 2004. *Water Desalting Planning Guide for Water Utilities.* New York: Wiley.

Cath, T.Y., D. Adams, and A.E. Childress. 2005. "Membrane Contactor Processes for Wastewater Reclamation in Space: II. Combined Direct Osmosis, Osmotic Distillation, and Membrane Distillation for Treatment of Metabolic Wastewater." *Journal of Membrane Science,* 257:1–2:111–119.

Cath, T.Y., J.E. Drewes, and C. Lundin. 2009. "A Novel Hybrid Forward Osmosis–Reverse Osmosis Process for Water Purification and Reuse, Using Impaired Water and Saline Water." Proc. AWWA 2009 Membrane Technology Conference, Memphis, Tenn.

Eke, J., A. Yusuf, A. Giwa, and A. Sodiq. 2020. "The Global Status of Desalination: An Assessment of Current Desalination Technologies, Plants and Capacities." *Desalination,* 495:114633.

GWI (Global Water Intelligence). 2024. DesalData. www.desaldata.com (accessed December 1, 2025).

Holloway, R.W., A.E. Childress, K.E. Dennett, and T.Y. Cath. 2007. "Forward Osmosis for Concentration of Concentration From Anaerobic Digester." *Water Research,* 41:17:4005–4014.

IDRA (International Desalination and Reuse Association). 2024. *Desalination & Reuse Handbook*. Danvers, MA: IDRA.

Ingo, L., S.J. Metz, G. Rexwinkel, and G.F. Versteeg. 2009. "Desalination via Supercritical Water: A New Approach for Desalination Without Brine Streams." Proc. International Desalination Association 2009 World Congress, Dubai, UAE.

Makabe, R., T. Ueyama, H. Sakai, and A. Tanioka. 2021. "Commercial Pressure Retarded Osmosis Systems for Seawater Desalination Plants." *Membranes (Basel)*, 11:1:69.

Mallevialle, J., Odendaal, P.E., and Wiesner, M.R. 1996. *Water Treatment Membrane Processes*. New York: McGraw Hill.

Mayo, R. 2009. *The Current Status of Desalination Systems in San Juan County, Washington*. San Juan County Health and Community Services. www.sanjuancountywa.gov/DocumentCenter/View/861/Current-Status-of-Desalination-Systems-in-San-Juan-County-PDF (accessed June 6, 2025).

McCutcheon, J.R., and M. Elimelech. 2006. "Influence of Concentrative and Dilutive Internal Concentration Polarization on Flux Behavior in Forward Osmosis." *Journal of Membrane Science*, 284:1–2:237–247.

Sahle-Demessie, E., A.A. Hassan, and A. El Badawy. 2019. "Bio-desalination of Brackish and Seawater Using Halophytic Algae." *Desalination*, 465:9:104–113.

United Nations. 2021. *The Second World Ocean Assessment Volume I*. New York: United Nations.

Wade, N.M. 2001. "Distillation Plant Development and Cost Update." *Desalination*, 136: 1–3:3–12.

Yang K.L., T.Y. Ying, S. Yiacourmi, C. Tsouris, and E.S. Vittoratos. 2001. "Electrosorption of Ions From Aqueous Solutions by Carbon Aerogel: An Electrical Double Layer Model." *Langmuir*, 17:6:1961–1969.

Ying, T.Y., K.L. Yang , S. Yiacoumi, and C. Tsouris. 2002. "Electrosorption of Ions From Aqueous Solutions by Nanostructured Carbon Aerogel." *Journal of Colloids and Interface Science*, 250:1:18–27.

Chapter 2

# Water Quality

*Second Edition Author: Greg Wetterau*
*First Edition Authors: Greg Wetterau, Shane Trussell*

## 2.1 SOURCE WATER QUALITY

Seawater quality in the open ocean typically does not vary more than 10 percent over time and location; however, large variability can be seen in partially isolated seawater bodies, such as bays, estuaries, and seas, where influences from freshwater flows mixing with seawater and evaporation have considerable impact on the water quality patterns. Figure 2-1 shows the worldwide surface salinity of oceans, as modeled by the Ocean Climate Laboratory of the National Oceanographic Data Center (World Ocean Atlas 2018). The figure presents salinity in practical salinity units (psu), with one psu roughly equivalent to 1,000 mg per liter (mg/L). (Practical salinity units are defined as the ratio of electrical conductivity at 15°C and one standard atmosphere, compared with a fixed concentration of potassium chloride solution [Shkvorets 2025]). While higher-than-average salinities are common in areas receiving little rainfall—such as the Red Sea, the Mediterranean Sea, and the Arabian (Persian) Gulf—seawater bodies surrounding the United States are often average or below average in salinity. Considerably lower salinities are seen in partially closed water bodies receiving high contributions from freshwater flows like San Francisco Bay, Tampa Bay, Chesapeake Bay, and most of the inland waterways on the Atlantic and Gulf coasts where seawater desalination facilities are either currently operational or have been actively evaluated. The Atlantic Ocean generally has a higher salinity than the Pacific due to differences in ocean currents and weather patterns.

Extensive monitoring of seawater quality was first documented by William Dittmar in 1884, after 4 years of water quality sampling over nearly 70,000 nautical mi. Dittmar's numbers were updated in a 1940 paper from the Scripps Institute of Oceanography in California (Lyman and Fleming 1940), with adjustments made based on more accurate knowledge of atomic weights but most values remaining within 1 percent of those reported originally by Dittmar. The values he derived can be compared in Table 2-1 against more

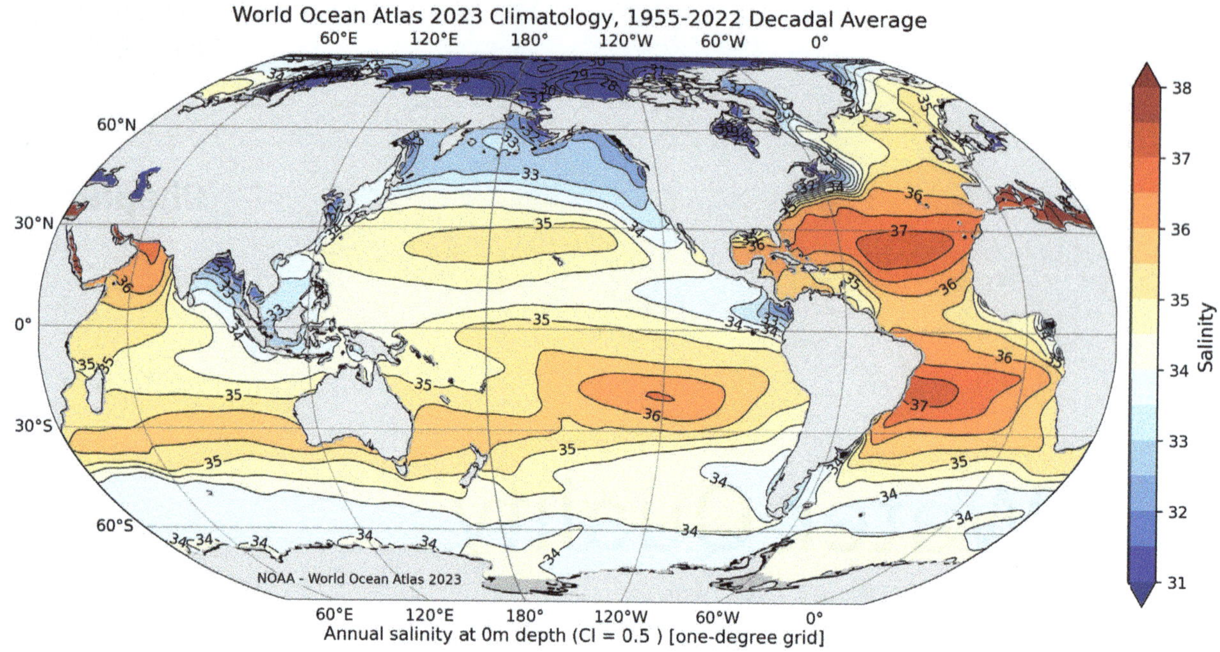


*Source:* Reagan et al. 2024

**Figure 2-1 Global sea-surface salinities**

**Table 2-1 Seawater mineral quality compared to conventional source waters**

| | Seawater | | | US Freshwater Supplies† | |
|---|---|---|---|---|---|
| Constituent | Average* mg/L | Carlsbad, Calif. mg/L | Aqaba, Jordan mg/L | Median mg/L | Upper Quartile mg/L |
| Chloride | 19,350 | 18,086 | 22,453 | 10 | 20 |
| Sodium | 10,710 | 9,474 | 12,374 | 30 | 80 |
| Sulfate | 2,690 | 2,412 | 3,143 | 30 | 75 |
| Magnesium | 1,304 | 1,298 | 1,507 | 10 | 20 |
| Calcium | 419 | 431 | 480 | 40 | 60 |
| Potassium | 390 | 387 | 437 | 2 | 4 |
| Alkalinity | 146 | 129 | 180 | 190 | 250 |
| Bromide | 70 | 78 | 80 | 0.02 | 0.06 |
| Boron | 4.4 | 4.3 | 5.5 | 0.08 | 0.2 |
| Fluoride | 1 | 0.9 | 1.2 | 0.2 | 0.4 |
| TDS | 35,079 | 34,500 | 41,510 | 330 | 480 |

*Source:* Greg Wetterau and Shane Trussell

* Dittmar 1884

† James Montgomery Consulting Engineers, Inc. 1985

recent analysis of isolated seawater bodies, such as the Red Sea (Aqaba, Jordan) or offshore analysis in the Pacific Ocean (Carlsbad, Calif.). The extremely high concentrations of dissolved minerals, many of which are orders of magnitude higher than typical freshwater supplies, demonstrate the challenge of treating seawater to a quality acceptable for potable water use.

While the concentrations of dissolved inorganic constituents are relatively consistent throughout the open ocean, concentrations of suspended material and colloids will vary drastically between locations and over time. As an example, the turbidity of seawater measured over 1 year of pilot testing at the Santa Cruz, Calif., desalination pilot varied by 2 to 30 ntu, with no significant storm events or red tide blooms experienced (Desormeaux et al. 2009). In contrast, a longer pilot conducted by West Basin Municipal Water District near Los Angeles never saw turbidity exceed 10 ntu, even during a severe red tide event (Filteau et al. 2008). Seawater turbidity can be impacted by suspended sediment from surface water runoff or the ocean floor or from phytoplankton, marine larvae, and other organic matter.

Variable levels of algal activity have also caused considerable issues with desalination plant performance at some locations, requiring enhanced pretreatment and increased cleaning of reverse osmosis (RO) membranes. A number of monitoring parameters have been proposed to help quantify the risk of biological fouling from variations in seawater quality. These include total organic carbon (TOC), total algal counts, transparent exopolymer particles (TEP), adenosine triphosphate (ATP), chlorophyll a, silt density index (SDI), and modified fouling index (MFI). Each of these is described briefly below.

### 2.1.1 TOC

TOC is a commonly used parameter in water treatment for characterizing overall organic content. The techniques outlined in *Standard Methods* for measuring TOC (methods 5310 B, C, and D) are problematic or inaccurate when used on seawater, requiring special equipment or modified testing methods for accurate results. Low-TOC water (2 mg/L or less) generally has low fouling potential for ultrafiltration (UF) and RO membranes, and higher-TOC water will require more extensive pretreatment. Variable levels of TOC can be caused by storm events, algal blooms, or changes in ocean currents.

### 2.1.2 Total Algal Counts/Algal Density

Total algal count or algal density is the direct measure of algal concentrations through filtration and quantifying the algal colonies retained on the test filter. The process is laborious and time consuming but provides a direct quantification of the algal cells present in the water. Flow cytometry can also be used as a more rapid means of quantifying algae in seawater. Algal concentrations in the open ocean can vary from less than 10 cells/mL to over 1,000 cells/mL during algal bloom events.

### 2.1.3 TEP

TEP monitoring has been utilized to quantify fouling potential from biological material without resorting to direct counting of microbial colonies. TEP are polysaccharide-rich, gel-like particles derived from biological material. They can be measured directly using spectrophotometric methods, giving instantaneous results and allowing for field measurement of both colloidal (c-TEP) and particulate (p-TEP) species. TEP concentrations are expressed as equivalent values of a standard polysaccharide, xanthan gum, as milligrams xanthan gum equivalents per liter (mg Xeq/L) (Nagata et al. 2021). TEP concentrations in the open ocean can vary from less than 0.1 to over 10 mg Xeq/L.

### 2.1.4 ATP

ATP is a compound found in all living material, associated with the transport of energy within cells. It can be measured in seawater using electrochemical impedance spectroscopy or bioluminescent monitoring, which can quantify extracellular (free) ATP or microbial

ATP to project active microbial biomass in water (Abushaban et al. 2019). Typical ATP levels in the open ocean have been recorded from less than 10 to over 1,400 ng/L. Recent advances in analytical measurements can provide cellular ATP (c-ATP) counts, which may be a better representation of c-ATP within active cells versus ATP, which contains active cells as well as inactive cells and ATP lysed from cells or otherwise in bulk liquid. Though c-ATP aims to provide a better representation of biological activity, it has not been fully proven to be effective in being monitored via online probes in real time at all levels. Grab sampling with ATP and c-ATP analysis is available with current technologies at a detection level above 100 picograms/mL (1 pg/mL = 1 ng/L).

### 2.1.5 Chlorophyll a

Chlorophyll a can be monitored using spectrophotometry as an indirect measure of algal density in seawater. Chlorophyll-a levels in the open ocean have been measured from less than 0.1 μg/L to over 10 μg/L during algal bloom events (Dhakal et al. 2019; Ozair et al. 2019).

### 2.1.6 SDI and MFI

SDI and MFI are common methods used for quantifying particulate fouling potential in seawater. The tests both involve the use of a 0.45-micron filter disc to filter the water for a fixed period of time at a fixed pressure. While the SDI test is more widely used in the industry, MFI has been adopted by an increasing number of utilities because of its better correlation with colloidal material concentrations and its ability to correct for temperature variability and permeability differences in test membranes. RO vendors generally require 15-min SDI values less than 5, with values less than 3 recommended. These targets can generally not be achieved in seawater extracted from open ocean intakes without pretreatment. Because of the fouling potential of raw seawater, the duration of the SDI/MFI test is often reduced to 3–5 min, generating values that can be used for site-specific characterization.

Beyond the variations in suspended solids and organic or biological content, localized variation is often seen in nutrient concentrations like nitrate, ammonia, and phosphate, impacted by surface runoff and the influence of human irrigation practices. Elevated hydrocarbon concentrations can also be seen in areas impacted by heavy shipping traffic or offshore oil production. These hydrocarbons can be detrimental to desalination membranes if not adequately addressed through pretreatment.

## 2.2 PRODUCT WATER QUALITY

The treatment processes used in seawater desalination result in a product water having very low mineral content, with the exception of a few constituents that are either insufficiently removed by RO membranes (such as boron and bromide) or are exceptionally high in the source water (such as sodium and chloride). These water quality characteristics result in unique water that may be further conditioned, treated, or possibly blended to produce a finished water that is acceptable for use in potable water systems. The key issues that must be addressed include concerns about health, product water stability, corrosion inhibition, compatibility with existing distribution networks, irrigation and industrial use, and general aesthetics.

The remaining sections of this chapter will provide a greater level of detail on each of these four topics, discussing the key water quality parameters associated with each and the methods commonly used to meet the treatment and water quality goals.

## 2.3 HEALTH CONCERNS

Issues related to public health in drinking water are regulated under the US Environmental Protection Agency (USEPA) as administered through state health and environmental protection departments. While the requirements for seawater desalination facilities are the same for all public drinking water systems, the key health concerns with desalinated product water differ somewhat due to the unique nature of the source water and the processes used in desalination. The key health issues can be divided into issues related to potentially high levels of specific minerals in the RO product, issues related to pathogen removal, and issues related to distribution system water quality.

### 2.3.1 Mineral Content of Product Water

Seawater RO (SWRO) membranes used in desalination produce high-quality water that meets water quality requirements for most regulated compounds. It is generally only boron that has a significant risk of exceeding current federal or state guidelines, and these guidelines are based primarily on issues other than public health, as will be discussed elsewhere in this manual. Similar to boron, the chloride concentrations produced by SWRO membranes may also be a concern for non-health related issues; however, these concentrations are unlikely to exceed state and federal guidelines for chloride. Health concerns have also been associated with low mineral-content water, with some studies suggesting insufficient levels of calcium and magnesium in drinking water may contribute to health risks for people who drink predominantly desalinated water (Shlezinger et al. 2018). The presence of some unregulated organic compounds in ocean water, such as algal toxins, has also caused some concern among state and federal regulatory agencies, with various studies done to document the removal effectiveness and prevalence of these compounds.

**Boron.** Boron is poorly removed by most RO membranes due to its low molecular weight and weak ionic charge. Typical removals vary from 40 to 90 percent at near neutral pH. Higher removals can be achieved at elevated pH, with greater than 99-percent removal reported at pH greater than 9.5 (Oo and Song 2009). With levels of boron in seawater typically around 4.4 mg/L, RO product from a single-pass treatment approach is typically on the order of 1–2 mg/L, depending on the temperature, flux, recovery, and specific membranes employed.

Animal tests with boron have shown adverse effects on the reproductive systems of rats and male dogs; however, the USEPA decided as recently as 2008 to not regulate boron with a maximum contaminant level (MCL) due to its low prevalence at levels of concern in US drinking water supplies (USEPA 2008b). The USEPA has established a long-term health advisory level for children of 2 mg/L and for adults at 5 mg/L based on health effects observed during animal testing. In addition, several states (California, Florida, Maine, Minnesota, New Hampshire, and Wisconsin) have adopted standards, guidelines, or notification levels for boron ranging from 0.6 to 1 mg/L (USEPA 2008a). Not all of these limits are driven by public health concerns. Non-health–based concerns associated with boron are discussed in Section 2.5. Beyond US guidelines, the World Health Organization (WHO) has set a health guideline for boron since 1993. This was set initially at 0.3 mg/L, was increased to 0.5 mg/L in 1998, and to 2.4 mg/L in 2009. The current WHO guideline is based on public health risks, rather than aesthetic or other non-health concerns (WHO 2009a).

**Chloride.** Chloride is generally not raised as a health issue for drinking water but is established as a secondary MCL (set at 250 mg/L), primarily for aesthetic concerns related to the taste of high chloride water. Chloride levels in raw seawater would certainly be fatal, if consumed in significant quantities, but due to the high removal achieved by RO membranes, RO product from a single-pass system ranges from 100 to 200 mg/L, depending on the temperature, flux, and specific membranes employed. Lower levels may be sought by some utilities based on aesthetic concerns.

**Hardness.** Calcium is generally added to desalinated water to provide stabilization and protect the product water from degradation within the storage, transmission, and distribution systems. Product water stability will be discussed later in this chapter. In addition to stabilization concerns, there are direct health concerns that may be linked to long-term consumption of water with insufficient levels of calcium and magnesium. According to the WHO, inadequate calcium is associated with increased risks of osteoporosis, kidney stones, colorectal cancer, hypertension, stroke, coronary artery disease, and obesity. Similarly, inadequate magnesium has been implicated in hypertension, coronary heart disease, type 2 diabetes, and metabolic syndrome (WHO 2009b).

A recent study in Israel found that the rate of ischemic heart disease increased with the introduction of desalinated water into drinking water supplies but did not show any link to increased diabetes or colorectal cancer (Shlezinger et al. 2018). The Israeli Ministry of Health has set a goal of restoring magnesium in drinking water supplies to the historic level of 30 mg/L; however, no specific water quality standards have been established toward this end, and existing desalination facilities within Israel do not utilize magnesium in their product water stabilization process. Calcium addition, which is far more common in desalinated product water, typically targets a total hardness above 30 mg/L. While there is still debate about potential health impacts of low-hardness water, it is the need for product water stabilization that typically dictates the addition of hardness.

**Nonregulated parameters.** A large number of emerging contaminants or contaminants of emerging concern are being studied in terms of their health effects and prevalence in US water supplies. While some of these contaminants can be found in seawater sources at concentration approaching those in freshwater supplies, most are considerably lower in seawater and are less of a concern to seawater desalination than they are to conventional water treatment facilities. As an example, a global survey of per- and polyfluoroalkyl substances (PFAS) in oceans found that levels of perfluorooctane sulfonate (PFOS) did not exceed 0.05 ng/L in the central Pacific Ocean (Yamashita et al. 2005), and levels in coastal Atlantic areas near the United States have not exceeded 0.1 ng/L (Zhang et al. 2019). These levels are significantly lower than both existing health advisory levels and levels common in freshwater bodies of the United States. Even if these concentrations reached a level of concern for public health, RO is recognized as a best available technology for removal of PFAs, and elevated levels would not be anticipated in the product water.

Compounds of unique concern to seawater desalination are currently unregulated toxins related to periodic algal blooms or red tide events. Two common algal toxins detected in US waters are found off the Pacific Coast and include domoic acid, the cause of amnesic shellfish poisoning, and saxitoxin, the cause of paralytic shellfish poisoning. The toxins are typically not found in the water column at concentrations that are considered toxic but have been found to accumulate in shellfish to levels that are toxic to both humans and other mammals. The algae that produce the toxins are readily removed by the pretreatment filters within a desalination plant and by the RO membranes; however, there is a concern that cells will break during the pretreatment filtration process and release more dissolved toxins into the water. Pilot studies in southern California (Carlsbad and West Basin) have demonstrated excellent removal of the toxins during large harmful algal blooms. In addition, a spiking of dissolved toxins at 1,000 times typical concentrations during pilot testing in Santa Cruz demonstrated greater than 3 log (>99.9%) rejection using kainic acid, a common surrogate for domoic acid.

### 2.3.2 Pathogen Removal

The USEPA Surface Water Treatment Rule (SWTR) requires removal and inactivation of viruses and *Giardia* cysts, as well as turbidity reduction, for surface water and groundwater under the direct influence of surface water (GWUDI). The Long Term 2 Enhanced Surface Water Treatment Rule (LT2ESWTR) sets removal and inactivation requirements for *Cryptosporidium*. Seawater is currently regulated as a surface water supply in the United States; however, the biological activity in seawater and the survival rate of common pathogens are considerably different in seawater than in natural freshwater bodies. The survival rates of *Escherichia coli*, *Giardia* cysts, and *Cryptosporidium* oocysts have been shown to be low in seawater when compared with freshwater systems. Nonetheless, these pathogens have been found to be present in seawater bodies directly impacted by wastewater and stormwater outfalls (Fayer et al. 1998).

Pathogen removal requirements established by the USEPA require a multiple barrier approach to provide both removal and inactivation of pathogens. The SWTR requires a minimum of 99.99 percent (4-log) reduction in viruses and 99.9 percent (3-log) reduction in *Giardia* through filtration and disinfection. The LT2ESWTR requires between 2-log and 5.5-log reduction in *Cryptosporidium*, depending on the prevalence of *Cryptosporidium* found in 2 years of source water monitoring and on the type of treatment employed. In addition to the federal requirements, some states include multiple barrier requirements for specific pathogens (*Giardia* and viruses) and may require higher removal for impaired water bodies where high concentrations of total coliforms have been measured in Watershed Sanitary Survey (WSS) monitoring. Table 2-2 summarizes the minimum and maximum reduction requirements for *Giardia*, *Cryptosporidium*, and viruses.

### 2.3.3 Distribution System Water Quality

Distribution system water quality concerns relate both to corrosion within the distribution system and to the formation of disinfection byproducts (DBPs). Corrosion issues, while relating to health concerns about lead and copper leaching from residential plumbing, also have aesthetic components and are discussed later in Section 2.4 on product water stability. This section will focus primarily on the impact of desalinated water on DBP formation. The formation of DBPs such as trihalomethanes (THMs) are impacted primarily by the presence of natural organic matter (NOM) when free chlorination is used for disinfection. Levels of NOM in desalinated product water are considerably lower than in most conventional treatment products. The formation of DBPs is therefore a concern primarily when desalinated seawater is blended in a distribution system with treated surface waters containing significant levels of NOM.

As noted in Table 2-1, seawater contains unusually high levels of bromide when compared with most freshwater sources in the United States, with the concentration of bromide roughly 3 orders of magnitude larger in seawater than in the highest quartile

**Table 2-2 Log reduction requirements**

| | Log Reduction Requirement | | |
|---|---|---|---|
| Pathogen | Minimum | Maximum | Criteria for Determination |
| Virus | 4 | 6 | Total coliforms* |
| *Giardia* | 3 | 5 | Total coliforms* |
| *Cryptosporidium* | 2 | 5.5 | *Cryptosporidium*,† total coliforms* |

* California Division of Drinking Water requires up to 2-log additional removal where high levels of total coliforms have been detected.

† The Long Term 2 Enhanced Surface Water Treatment Rule requires up to 5.5 log *Cryptosporidium* removal based on bin classification from *Cryptosporidium* monitoring.

of freshwater supplies. Bromide removal with SWRO membranes is comparable to the removal of chlorides; however, the level of bromide that results from seawater treatment in a single-pass system is still an order of magnitude larger than in most freshwater supplies (0.5–0.8 mg/L under typical operating conditions). High bromide levels have been associated with the formation of brominated DBPs. Chlorine added for disinfection can oxidize bromide to bromine; the bromine may then react with NOM to form brominated DBPs. Brominated DBPs are a concern both due to an increased health risk compared with chlorinated DBPs and because brominated compounds increase the mass-based concentration more than chlorinated DBPs due to the molecular weight of bromide, which is approximately double that of chloride.

Brominated DBPs can be expected to form in zones where desalted water blends with other supplies when free chlorine is used for disinfection residual maintenance. DBP formation tests are commonly recommended with blends of the desalinated water and local treated surface water to assess if this is a potential issue. Previous studies of desalinated water disinfection have also shown that high levels of bromide can have adverse impacts on the stability of chloramine residual in distribution systems (Heeb et al. 2014; Deborde and von Gunten 2008). Currently regulated haloacetic acids (HAAs) for DBPs consist of HAA5, which includes dibromoacetic acid, dichloroacetic acid, monobromoacetic acid, monochloroacetic acid, and trichloroacetic acid. The proposed USEPA M/DBP rule may add four more HAAs—bromochloroacetic acid, bromodichloroacetic acid, chlorodibromoacetic acid, and tribromoacetic acid—to change the name to HAA9. These four HAAs all have impacts from bromine.

## 2.4 PRODUCT WATER STABILITY

Seawater desalination results in a product water quality that is characterized by low hardness (calcium and magnesium), low alkalinity, and relatively high chloride. The hardness and alkalinity are so low that desalination facilities typically condition the product water to achieve a slightly positive Langelier saturation index (LSI) to control corrosion in the distribution system and consumer plumbing. Soft water supplies are well known to have a higher potential to create corrosion-related apparent color (red water) problems when conveyed in iron pipe and to cause leaching of lead and copper in home plumbing, which can increase health risks and violate requirements of the Lead and Copper Rule. High chloride content may also exacerbate these problems, increasing the corrosivity of the water. Cast-iron or ductile-iron pipes, either unlined or lined with older epoxy coatings, are susceptible to iron release and red water complaints if the softened water is not appropriately conditioned for corrosion protection.

The LSI, an index first defined by Wilfred Langelier in 1936 (Langelier 1936), has become the most widely used measure of product water aggressivity, and a positive LSI (>0) is used as a target for many desalination posttreatment facilities. Other measures of product water aggressivity include the calcium carbonate precipitation potential (CCPP), which is often kept at 4–10 mg/L, and the aggressiveness index (AI), which is often kept above 12. Calcium saturation indices do not imply knowledge of the kinetics (speed) or precipitation or dissolution, and generally phosphates will increase chemical precipitation tendency of calcium carbonates. The adjustment of calcium carbonate saturation, however, is not uniformly practiced nor is it always successful in meeting corrosion-control objectives. Other common approaches for mitigation of corrosive effects from soft water can include pH adjustment and addition of orthophosphate. It is generally most important, however, to avoid significant variations in either pH or calcium carbonate saturation to prevent problems with pipeline corrosion or customer complaints from cloudy or red water.

## 2.5 IRRIGATION AND INDUSTRIAL USE CONCERNS

While desalinated product water from seawater supplies has a low risk of exceeding health-related water quality goals, the levels of boron, chloride, and sodium in the product of single-pass RO system could create significant problems for irrigation and industrial users. The primary issues related to these types of users are discussed in the next few sections.

### 2.5.1 Irrigation and Agricultural Use

High levels of chloride, sodium, and boron are known to impact numerous different ornamental plants, citrus trees, and other common agricultural crops. Although the total dissolved solids (TDS) concentration is low in desalinated ocean water, the concentrations of chloride and sodium comprise a majority of the TDS. Chloride and sodium toxicity have been more commonly observed than boron toxicity to date because of the increased use of reclaimed water for irrigation purposes, particularly in the arid southwest. Some of the better-known cases of chloride toxicity have occurred where irrigation of redwood and avocado trees was converted from using local water supplies to recycled water, resulting in significant impacts on the health of these species. The most common symptom of chloride, sodium, and boron toxicity is tip burn, but chloride or sodium toxicity can also result in tattered leaves, reduced leaf size, reduced growth rate, yellowing of leaves in conifers, and in extreme cases, plant death. Similar to boron, chloride and sodium accumulate in the older leaves, and these leaves typically exhibit the symptoms first. Figure 2-2 shows an example of chloride toxicity observed in southern California and caused by irrigation using reclaimed water with high chloride content.

The sodium adsorption ratio (SAR) is a parameter used in agriculture to ensure safe water quality for irrigation supplies. The SAR is calculated as the sodium concentration (in milliequivalents) divided by the square root of the average between the calcium and magnesium concentrations. Water with an SAR of 13 or higher will contribute to poor soil conditions for plant growth, and an SAR of 5 or less is considered ideal. Desalinated

*Source:* Nelda Matheny

**Figure 2-2 Boston ivy with tip burn from chloride**

*Source:* Nelda Matheny

**Figure 2-3 Boron toxicity on camphor**

seawater typically has an SAR around 20 before adding calcium for product water stabilization.

Concerns with boron may arise if there is a significant difference between the boron content of the traditional water supply source or sources and the desalinated water, and if chloride- and boron-sensitive plants and crops are present in the service area. Figure 2-3 shows boron toxicity.

Boron is essential for plant growth; however, concentrations greater than 2 mg/L are unhealthy for most plants, and a boron concentration greater than 1 mg/L may have an impact on sensitive plant species (for a discussion on impacts to human health, see Section 2.3.1). If the boron concentration of water used for irrigation is maintained below 0.5 mg/L, boron toxicity should not result in significant impacts on plant appearance, particularly for water maintained at above neutral pH. Boron toxicity is known to increase with decreasing pH and decreasing calcium levels. Water with boron concentration of 0.75–1.0 mg/L may impact some trees, plants, and ornamentals that are commonly used for landscapes and residences in southern coastal areas of the United States where seawater desalination is being considered. Some of these plants include gardenia, crape myrtle, camellia, figs, giant birds of paradise, heavenly bamboo, hydrangea, lemons, lily of the Nile, oranges, peaches, philodendron, photinia, pink trumpet vine, southern magnolia, violet trumpet vine, dwarf pittosporum, xylosma, and some roses (Gupta et al. 2014; HortScience 2005).

Yermiyahu et al. (2007) reported that the seawater desalination plant in Ashkelon, Israel (73 mgd), is producing desalinated water with a boron level of 0.2–0.3 mg/L to meet Israeli recommendations for domestic and agricultural usage. A large portion of the water produced by the Ashkelon plant is used for irrigation of citrus crops, which are sensitive to boron levels in the water of more than 0.75 mg/L. The main rationale for producing a water with half this boron level was to reduce the overall boron level of existing surface water supplies of naturally high boron content with which this desalinated water is blended. In contrast, the Carlsbad desalination facility in California targets a boron level of 1 mg/L for 95 percent of samples, while the Monterey, Calif., desalination facility is planned to target a maximum boron concentration of 0.7 mg/L.

### 2.5.2 Industrial Use

The level of chloride in the treated water may also be an important consideration for industries located within the service area of the desalination plant. As mentioned above, the maximum chloride concentration in desalinated seawater treated using only a single-pass RO system is significantly higher than the concentrations observed in conventional water supplies (typically 150–220 mg/L versus less than 100 mg/L). It is known that some industries (i.e., petrochemical, power generation, pharmaceutical, breweries, and semiconductor wafer manufacturing) use high-purity water that has chloride levels significantly lower than 100 mg/L. If these types of industries are large water users in the desalination plant service area, product water quality targets in terms of chloride, TDS, and other compounds may need to be adjusted to accommodate their requirements. An example is Contra Costa Water District in Northern California, which has set a water quality objective of 100 mg/L chloride based on the requirements of the industries served in the desalination plant service area. Trinidad and Tobago Water and Sewerage Authority set a chloride objective of 80 mg/L for the 40-mgd Point Lisas seawater desalination facility based on the requirements of the industries served there. The Point Lisas plant is located in an industrial park where over 80 percent of its water is used by local industries.

Many of the industrial plants that have stringent requirements in terms of TDS, chloride, or other compounds are often equipped with water purification facilities, and it is often more cost effective for these installations to be upgraded or modified rather than to impose a significant incremental cost increase for a more elaborate desalination system to the general population. This is especially true when industries with very stringent product water quality requirements constitute a relatively small portion of the water demand in the plant service area. Many industrial uses may have standards that are less stringent than drinking water. Chilean desalination facilities using desalinated water for mining supplies target chloride and TDS levels up to 500 and 850 mg/L, respectively, driven more by corrosion concerns in the extensive transmission lines than by water quality needs for the end uses. These allowable levels are higher than Chilean desalination facilities serving primarily potable uses.

Industrial use of cooling water is often desired with three to six cycles of concentration (CoC) where the treated water is recycled by the desired CoCs. This can lead to limitations for water use due to chloride levels of up to 2000 mg/L that can impact corrosion or discharge requirements.

### 2.5.3 Addressing Irrigation and Industrial Use Concerns

To address the issues with boron, sodium, and chloride discussed above, several seawater desalination plants worldwide have been designed with RO systems that have a partial or full second pass, where a portion of the permeate produced by the first-pass RO system undergoes additional RO treatment. By using a partial or full second pass, boron and chloride levels in the finished product water can be targeted to specific needs of irrigation and industrial users in the area served by the desalination facility.

Examples of facilities that have a partial or full RO second pass include Tampa, Fla. (25 million gal per day [mgd] or 95 megaL per day [ML/d], 50-percent second RO pass); Trinidad and Tobago (36 mgd or 136 ML/d, 100-percent second RO pass); Fukuoka, Japan (13 mgd or 49 ML/d, 100-percent second RO pass); and Tuas, Singapore (36 mgd or 136 ML/d, 100-percent second RO pass). In contrast, most of the seawater desalination plants in Spain, Cyprus, and Malta providing water for both human consumption and agricultural irrigation for grain crops and vegetables have single-pass SWRO systems only. Most municipal SWRO treatment plants in the Caribbean are also built with a single-pass systems.

## 2.6 GENERAL AESTHETIC CONCERNS

Although aesthetics is typically considered better for desalinated product water than most conventional water supplies, some concerns should be considered when variability is possible in the temperature or taste and odor (T&O) of the water delivered to residential customers. These two issues are discussed briefly below.

### 2.6.1 Temperature

Ambient ocean water temperature is typically lower than or equal to that of most conventional surface water sources. Membrane desalination processes do not result in significant increases in product water temperature; therefore, desalinated seawater is typically of temperature comparable to that of other water sources. However, in the case of seawater desalination plants that are colocated with power plants, where cooling water from a power plant is used as a source for desalination, the product water may be 2–5°C (3.6–9°F) higher than the ambient ocean water temperature. Therefore, in this case, the temperature of the desalinated water may be higher than that of the existing water sources, which may have an impact on water quality aesthetics. The temperature aspect of desalinated seawater should be evaluated on a case-by-case basis. For example, experience at the Tampa Bay Water seawater desalination plant and environmental impact reports of the colocated Carlsbad and Huntington Beach seawater desalination plants indicate that the desalinated water temperature is well within the temperature of all other water sources in the desalination plant's service area.

### 2.6.2 T&O

Depending on how the desalinated water is introduced to the distribution system, water supply to some consumers may be converted entirely from the conventional water source to the new desalinated water. If these consumers remain on the new supply 100 percent of the time, following a transition period, the number of T&O complaints would not be expected to be significant as a result of the conversion to desalinated water. It is common, however, that on days of peak water demand or days when portions of a desalination plant are down for maintenance, a water utility may need to supplement their desalinated water supply with conventional water supplies, creating a risk of T&O complaints. It is important to develop an appropriately designed connection point and distribution system infrastructure that prevents consumers from transitioning between the two supplies on a routine basis. The number of T&O complaints would be expected to increase proportionally with the frequency with which consumers are transitioned from one supply to another. This means that each time conventional water supplies are required to supplement the desalinated water that there would be two sensitive periods where T&O complaints would be likely to occur. Any significant transition in water supply will be noted by consumers, and T&O complaints will become a significant issue if the delivered water source is frequently changed. This issue can be addressed by ensuring adequate blending and mixing with a local supply before distribution, which may allow for providing a consistent water supply. Such an approach was taken for the Sand City coastal desalination facility near Monterey, Calif., where up to 50 percent of distribution system water is blended with the desalination product to provide a consistent water quality within the system.

## REFERENCES

Abushaban, A., S.G. Salinas-Rodriguez, M.N. Mangal, S. Mondal, S.A. Goueli, A. Knezev, J.S. Vrouwenvelder, J.C. Schippers, and M.D. Kennedy. 2019. "ATP Measurement in

Seawater Reverse Osmosis: Eliminating Seawater Matrix Effects Using a Filtration-Based Method." *Desalination*, 453:1–9.

Deborde, M., and U. von Gunten. 2008. "Reactions of Chlorine with Inorganic and Organic Compounds During Water Treatment—Kinetics and Mechanisms: A Critical Review." *Water Research*, 42:1–2:13–51.

Dhakal, N., S.G. Salinas-Rodriguez, F. Knops, J.C. Schippers, and M.D. Kennedy. 2019. "Algal Blooms and Fouling Potential in Membrane Based Desalination Systems." Proc. International Desalination Association 2019 World Congress on Desalination and Water Reuse, Dubai, UAE.

Dittmar, W. 1884. "Report on Researches into the Composition of Ocean Water Collected by HMS Challenger, During the Year 1873–1876." *Report on the Scientific Results of the Voyage of HMS: Vol. I, Physics and Chemistry*. London: Longman & Co.

Fayer, R., T.K. Graczyk, E.J. Lewis, J.M. Trout, and C.A. Parley. 1998. "Survival of Infectious *Cryptosporidium parvum* oocysts in Seawater and Eastern Oysters (*Crassostrea virginica*) in the Chesapeake Bay." *Applied and Environmental Microbiology*, 64:3:1070–1074.

Gupta, U., P. Verma, and H. Solanki. 2014. "Impact of High Boron Concentration on Plants." *Biolife*, 2:2:610–614.

Heeb, M.B., J. Criquet, S.G. Zimmermann-Steffens, and U. von Gunten. 2014. "Oxidative Treatment of Bromide-containing Waters: Formation of Bromine and Its Reactions with Inorganic and Organic Compounds—A Critical Review." *Water Research*, 48:15–42.

James M. Montgomery Consulting Engineers, Inc. 1985. *Water Treatment Principles and Design*. New York: Wiley.

Langelier, W.F. 1936. "The Analytical Control of Anti-Corrosion Water Treatment." *Journal AWWA*, 28:10:1500–1521.

Lyman, J., and R.H. Fleming. 1940. "Composition of Seawater." *Journal of Marine Research*, 3:2:134–146.

Nagata, T., Y. Yamada, and H. Fukuda. 2021. "Transparent Exopolymer Particles in Deep Oceans: Synthesis and Future Challenges." *Gels*, 7:3:75.

Oo, M.H., and L. Song. 2009. "Effect of pH and Ionic Strength on Boron Removal by RO Membranes." *Desalination*, 246:1–3:605–612.

Ozair, G., S.A. Al-Zahrany, and A.A. Al-Shubaili. 2019. "Microbial Diversity in Seawater Near Marafiq Desalination Plants Ratifies Nonthreatening Impact on Marine Ecosystem." Proc. International Desalination Association 2019 World Congress on Desalination and Water Reuse, Dubai, UAE.

Reagan, J.R., D. Seidov, Z. Wang, D. Dukhovskoy, T.P. Boyer, R.A. Locarnini, O.K. Baranova, A.V. Mishonov, H.E. Garcia, C. Bouchard, S.L. Cross, and C.R. Paver. 2024. *World Ocean Atlas 2023, Volume 2: Salinity*. Silver Spring, MD: NOAA/NESDIS.

Shkvorets, I. 2025. Practical Salinity Scale–1978. Salinometry. https://salinometry.com/pss-78/ (accessed October 15, 2025).

Shlezinger, M., Y. Amitai, A. Akriv, H. Gabay, M. Schechter, and M. Leventer-Roberts. 2018. "Association Between Exposure to Desalinated Sea Water and Ischemic Heart Disease, Diabetes Mellitus and Colorectal Cancer; A Population-based Study in Israel." *Environmental Research*, 166:620–627.

USEPA (United States Environmental Protection Agency). 2008a. Drinking Water Health Advisory for Boron, US Environmental Protection Agency, Office of Water, Document No. 822-R-08-013. Washington: USEPA.

USEPA. 2008b. "Drinking Water: Regulatory Determinations Regarding Contaminants on the Second Drinking Water Contaminant Candidate List." Federal Register Vol. 73, No. 147 (July 30, 2008): 44251–44261. www.govinfo.gov/content/pkg/FR-2008-07-30/pdf/E8-17463.pdf (accessed December 1, 2025).

WHO (World Health Organization). 2009a. *Boron in Drinking Water: Background Document for Development of WHO Guidelines for Drinking-water Quality*. Geneva: WHO.

WHO. 2009b. *Calcium and Magnesium in Drinking-water: Public Health Significance*. Geneva: WHO.

Yamashita, N., K. Kannan, S. Taniyasu, Y. Horii, G. Petrick, and T. Gamo. 2005. "A Global Survey of Perfluorinated Acids in Oceans." *Marine Pollution Bulletin*, 51:8–12:658–668.

Yermiyahu, U., A. Tal, A. Ben-Gal, A. Bar-Tal, J. Tarchitzky, and O. Lahav. 2007. "Environmental Science: Rethinking Desalinated Water Quality and Agriculture." *Science*, 318:5852:920–921.

Zhang, X., R. Lohmann, and E.M. Sunderland. 2019. "Poly- and Perfluoroalkyl Substances in Seawater and Plankton From the Northwestern Atlantic Margin." *Environmental Science & Technology*, 53:21:12348–12356.

# Chapter 3

# Treatment Approaches

*Second Edition Authors: Val S. Frenkel, Giancarlo Barassi*
*First Edition Authors: Larry VandeVenter, Robert Huehmer, Val S. Frenkel*

## 3.1 PRETREATMENT

Historically, effective pretreatment has been the most challenging issue confronting users of seawater reverse osmosis (SWRO) desalination systems. Pretreatment of seawater is normally necessary before its application to reverse osmosis (RO) membranes to manage potential foulants such as particulate, colloidal, inorganic/organic, and biological material that can potentially scale or foul the membranes. Membrane fouling can lead to increased feed pressure requirements, increased energy consumption, reduced permeate production, reduced permeate quality, increased chemical cleaning, and irreversible fouling of the membranes. Pretreatment requirements for seawater desalination facilities are determined by the type and location of seawater intake as well as local influent water quality and can include chemical addition, clarification, dissolved air flotation (DAF), media filtration, and/or membrane filtration. Chemical addition in the pretreatment process may include coagulants and filtration aids, as well as acid and/or scale inhibitor, which are added to reduce scaling of sparingly soluble salts on the RO membrane surface. Disinfection, either continuous or intermittent (i.e., shock chlorination), is often included in the pretreatment process to mitigate biological fouling, and this then requires a dechlorination step ahead of the RO process to prevent damaging the membranes. This section presents common approaches to SWRO pretreatment, RO design parameters, disinfection, posttreatment, energy recovery, corrosion, and materials of construction.

### 3.1.1 Quality of Pretreated Water

SWRO desalination systems use semipermeable membranes to produce a high-quality permeate and a waste stream called concentrate or brine. Because RO membranes generally reject high percentages of salts (>99 percent) in normalized conditions, salts accumulate on the feed side of the membrane, resulting in a concentrate. As the concentration approaches the solubility limits of sparingly soluble salts (e.g., calcium sulfate, barium sulfate, strontium sulfate, or calcium carbonate), precipitation of these salts onto the membrane surface may occur. This precipitation is referred to as scaling. Scaling increases the feed pressure requirements to move water through the membranes, which results in higher operating costs, increased salt concentrations in the permeate stream, increased cleaning requirements, and potential membrane damage. Certain scaling may prove to be irreversible.

SWRO system operators can control the concentration of sparingly soluble salts in the concentrate stream by controlling the system recovery, where a reduction in recovery lowers salt concentrations in the concentrate stream and thereby lowers the potential for membrane scaling. Another approach to minimize scaling is chemical conditioning of the feed water using scale inhibitors and dispersants. These chemicals can prevent or delay precipitation (even when some sparingly soluble species are supersaturated). Natural organic matter in seawater is believed to provide some level of natural scale inhibition (Chave and Suess 1970). As a result, many seawater installations have operated with no scale inhibitors at recoveries on the order of 50 percent with some sparingly soluble salt concentrations exceeding equilibrium conditions by as much as 300 percent. The ultimate recovery of SWRO units is typically determined by salinity of the source water, operating pressure, and salt passage. Detailed computer simulations can be performed to project system recovery based on water quality and operating conditions.

The primary indicator widely utilized by the industry for assessing SWRO pretreatment effectiveness is the silt density index (SDI). This simple analytical technique provides a qualitative measure of the fouling potential of the tested water. RO membrane manufacturers typically recommend SDI values less than 4.0, while a value of less than 3.0 is preferable. The SDI is not a perfect indicator of potential fouling, but it is still the current industry standard for indirect measurement of particulate and silt content in the source seawater. In general, lower SDI measurements indicate lower fouling potential that can result in longer intervals between membrane cleanings. Subjecting SWRO membranes to the rigors of chemical cleaning generally lowers the membrane life expectancy and increases the overall operations and maintenance costs of the system. A more extensive discussion on water quality parameters in SWRO is included in Chapter 2.

### 3.1.2 Pretreatment Technologies

The first commercial SWRO plant is generally considered to have been installed in Bermuda in 1974. By 1981, the United States was installing a similar capacity SWRO systems as Saudi Arabia, and the application of the technology, at least with seawater, was still small and developing. In the early 1980s, places such as Bahrain and Malta pushed the capacity of SWRO, and single train capacities reached 5 million gal per day (mgd) and then 10 mgd. There are currently thousands of SWRO plants operating worldwide with the largest facilities producing over 150 mgd. Multiple facilities producing 1 million $m^3/d$ (264 mgd) are currently in various stages of development, primarily in Middle Eastern countries. There are three general types of seawater intakes used in desalination facilities: open ocean intakes, beach wells, and infiltration galleries. The most commonly used intake approach is the open ocean intake, although beach well supplies have been widely used in the Caribbean and elsewhere and are now being viewed positively by regulators in California. A detailed discussion of intake alternatives is included in Chapter 4 of this manual.

Generally, seawater drawn directly from open intakes requires more robust pretreatment than seawater derived from beach wells or infiltration systems, so pretreatment for SWRO applications can generally be divided into two categories in relation to the feed water supply system: subterranean or subsurface intakes and surface supply intakes. The pretreatment requirements vary greatly as a function of which supply system is utilized. The following sections describe pretreatment approaches for both categories of intake facilities.

**Pretreatment for subterranean intake facilities.** Pretreatment in beach well applications, the most common form of subsurface intake, is typically limited to chemical addition for scale inhibition and cartridge filtration upstream of the RO membranes. Limited pretreatment is generally required as a result of low levels of particulate, colloidal, inorganic/organic, and biological material after the seawater has been prefiltered through the sand of the seafloor or beach. Additionally, naturally filtered seawater generally has relatively constant physical characteristics like temperature, turbidity, and microbial content. However, in cases where beach wells collect water from alluvial coastal aquifers, source waters may contain iron and manganese in quantities that could require pretreatment with green sand filters or other oxidation/filtration processes. Little information is available on the reliability of water quality from infiltration galleries because of their limited use in full-scale desalination facilities. A 13-mgd infiltration gallery has been in successful operation in Fukuoka, Japan, since 2005 (JICA 2022). In addition, preliminary testing at a pilot facility in Long Beach, Calif., has suggested that an infiltration gallery with a minimum 5 ft (1.5 m) of cover could provide a source water with a turbidity less than 1.0 ntu (Cheng et al. 2010).

**Pretreatment for surface supply intake facilities.** Because open seawater intakes are exposed to the surrounding conditions in the adjacent ocean water body, potential water quality issues that could substantially affect the ultimate plant cost are listed below. These parameters should be considered as indications that increased pretreatment is required to ensure stable SWRO operation:

- high turbidity greater than 20 ntu;
- occurrences of significant red tides or algae blooms;
- potential for high levels of pathogens;
- temperature of the raw water;
- moderate-to-high levels of total organic carbon (TOC) (e.g., Tampa Bay Seawater Desalinization Plant had TOC as high as 12 mg/L);
- seawater quality excursions caused by hurricanes or other severe storm events; and
- fat, oil, grease, and other hydrocarbon presence in the seawater.

Pretreatment system processes for open intake–derived seawater must be robust enough to handle expected worst-case variations in water quality and provide low levels of SDI after pretreatment for stable operation of the RO process.

Conventional pretreatment unit operations for open-intake SWRO plants have consisted of the following processes, each of which is described in Table 3-1:

- chlorination (shock or continuous),
- coagulation and flocculation,
- clarification (typically sedimentation),
- filtration (typically media filtration),
- chemical dosage for scale inhibition,
- cartridge filtration, and
- dechlorination.

**Table 3-1 Seawater reverse osmosis (SWRO) pretreatment components for surface seawater sources**

| Pretreatment Component | Description | Discussion |
|---|---|---|
| **Chlorination** | Chlorination is frequently used for disinfecting the intake and pretreatment system to mitigate biofouling in the downstream reverse osmosis (RO). Historically, continuous chlorination was used at levels up to 5 mg/L where cellulose acetate membranes were employed. For chlorine sensitive thin-film composite (TFC) membranes, the use of intermittent shock chlorination with downstream chlorine quenching has become more common for controlling biofilm growth. | It was believed that continuous chlorination was necessary to prevent RO biofouling. Chlorination of naturally occurring humic and fulvic acids creates high concentrations of assimilable organic carbon, which is currently known to be a principal player in the RO biofouling process. Intermittent shock chlorination has shown to be an improvement in many plants, while some have totally eliminated pretreatment chlorination with successful results. |
| **Coagulation/ flocculation** | Coagulation and flocculation are used to destabilize the suspended and colloidal material from the raw seawater. The most common coagulants include ferric salts such as ferric chloride and ferric sulfate (typically used at levels of 3–10 mg/L as Fe). Mechanical flocculation has been used successfully where additional mixing energy is required and/or downstream clarification is provided. In-line coagulation with only hydraulic flocculation is commonly used in treating warmer water (temperatures greater than 15°C or 59°F) with low fouling potential and no downstream clarification step. | Highly variable water quality may require coagulation/flocculation unit processes. Sufficient mixing is critical, especially when only in-line coagulation is followed by direct filtration (no sedimentation). |
| **Sedimentation** | Clarification can be applied by sedimentation using conventional clarifiers or high-rate clarifiers. Alternative clarification approaches have seen increased use and are discussed in Table 3-2. | Conventional sedimentation typically operates at 0.5 gpm/ft$^2$ (1.2 m/h) requiring a large footprint. High-rate clarifiers, such as lamella-type tube and plate settlers, require a smaller footprint and have been used at loading rates of 1.0–3.0 gpm/ft$^2$ (2.4–7.3 m/h) and 3.0–5.0 gpm/ft$^2$ (7.3–12.2 m/h), respectively. |
| **Filtration** | Media filtration may involve mono or dual media sand and anthracite or mixed media sand, anthracite, and garnet as media. Both single- and two-stage systems are common in SWRO, as are both pressure and gravity filters. Typical loading rates are 2–6 gpm/ft$^2$ (5–15 m/h). | Plants that use in-line coagulation may use two-stage filtration to address the higher solids loading from nonclarified water. Silt density index (SDI) goals of less than 3.0 units are generally desired with sufficient design in the coagulation and filtration processes. An extended filter-to-waste step after backwash is beneficial in diverting suboptimum filtered water from a clean filter back to the head of the plant and maintaining low SDI filtered water to the RO membranes. |
| **Chemical addition** | Sulfuric acid may be used to depress the pH to prevent calcium carbonate scaling. Scale inhibitors, with or without acid addition, have also been applied to prevent or delay sparingly soluble sulfate salts from precipitating. Hydrochloric acid can be used to reduce sulfate addition. | Acid addition has been used successfully at many plants; however, many of the newer plants are employing only scale inhibitors to reduce chemical usage. Some plants continue using combination of both acid and scale inhibitors. |

*(continued)*

**Table 3-1 Seawater reverse osmosis (SWRO) pretreatment components for surface seawater sources (*Continued*)**

| Pretreatment Component | Description | Discussion |
|---|---|---|
| **Cartridge filtration** | Cartridge filters are used as the last line of defense against particles reaching the RO membrane surface. However, the major purpose of the cartridge filters is a safety barrier to protect RO membranes from unwanted events when pretreatment is performing suboptimally. Typically, 5-μ filters are used; occasionally 1- to 3-μ are used, especially during initial commissioning. | When the coagulation/filtration processes underperform, the cartridge filters incur high loadings requiring frequent replacement. Iron deposits and biofouling are frequent complaints in a poor performing plant. |
| **Dechlorination** | RO membranes and TFC specifically are susceptible to chlorine oxidation, and therefore, all chlorine must be removed from the pretreated water. Sodium bisulfite (SBS) is the most commonly used dechlorinating agent. | Rapid biofouling occurs immediately following dechlorination in plants with continuous chlorination and high organic content. Additionally, reduction of ferric salts can create catalyzed chlorine oxidation. SBS alone is not problematic and has shown to have biostatic properties. |

*Source:* Larry VandeVenter

Depending on the source water quality, additional or alternative pretreatment processes have been implemented, such as filtration using diatomaceous earth and granular activated carbon (GAC). GAC is used most often in Arabian Gulf applications to remove oil and grease that may be present in the feedwater. In addition, several alternative approaches to SWRO pretreatment have seen increasing use over the past decade or have seen increasing consideration for adoption. Some of the following approaches are discussed further in Table 3-2:

- DAF in place of or in addition to sedimentation,
- membrane ultrafiltration (UF) and microfiltration (MF) in place of clarification and media filters,
- upflow solids contact clarification (Pulsator, SuperPulsator) in place of coagulation/flocculation, and
- microsand-enhanced settling (MES, Actiflo, or others) in place of conventional sedimentation.

A further description of pretreatment processes used in desalination can be found in AWWA Manual M46, *Reverse Osmosis and Nanofiltration*.

**Pretreatment at recently installed SWRO plants.** Outside the United States, SWRO has typically employed direct filtration as part of the pretreatment process. Single- or two-stage direct filtration can only be used when the raw water turbidity is relatively low. Water with higher levels of turbidity or algae often requires clarification before filtration. For example, a large SWRO plant in Trinidad is operating with tube settler clarification and single-stage filtration; this pretreatment regimen was considered necessary due to raw water turbidity excursions as high as 35 ntu where the typical turbidity is generally less than 10 ntu. The design loading rate of the tube settlers is 1.9 gal per min per $ft^2$ ($gpm/ft^2$) or 4.6 m per h (m/h), and the filters are designed at 6.5 $gpm/ft^2$ (15.9 m/h).

In the United States, the largest SWRO facility on the east coast (Tampa Bay, Fla.; 25.0 mgd/95 ML/d) initially used direct filtration with two-stage continuously backwashing upflow filters. However, during preliminary operation of this plant, there were severe problems with the pretreatment process because of poor feedwater quality, including growth of mussels and problems with mussel shell fragments. Corrective measures included conversion of the original two-stage sand filtration system into single-stage filters

**Table 3-2 Seawater reverse osmosis (SWRO) pretreatment nonconventional technologies for surface seawater sources**

| Pretreatment Component | Description | Discussion |
|---|---|---|
| **Dissolved air flotation (DAF)** | DAF can be used for clarification upstream of conventional filtration as a standalone clarification process or can be stacked over filtration. DAF can also be applied upstream of the low-pressure membranes (microfiltration/ultrafiltration [MF/UF]).<br>DAF uses microbubbles to float coagulated solids, colloidal material, and algae. | DAF operates at 6–12 gpm/ft$^2$ (15–29 m/h), significantly higher loading rates than conventional sedimentation or plate settlers, and has a smaller footprint. Pilot data have shown DAF can improve pretreated water quality, especially in removal of algal content (red tide). DAF has been employed with SWRO in Singapore and Morocco. While DAF typically uses polymers, it can be used without polymers at reduced efficiencies. |
| **Membrane filtration** | Membrane filtration uses MF or UF to replace conventional pretreatment or to replace just the filtration process following some form of coagulation. MF/UF systems are significantly more compact than media filters and can allow a substantial footprint savings when flocculation and clarification are not used. | Multiple vendors of MF and UF are available in a variety of formats, either pressure or submerged. Pilot studies have shown reduced or eliminated coagulant dosage using membrane filtration and enhanced pretreated water quality, with silt density index values generally well under 3.0. Membrane filtration has been successfully employed with SWRO in many locations, including Australia, South America, the Caribbean, and the Mediterranean. |
| **Upflow solids contact clarification** | Upflow solids contact clarifiers combine coagulation and flocculation within a pulsed blanket and are widely utilized in freshwater treatment. They combine coagulation and flocculation within a pulsed blanket. | Upflow solids contact clarifiers operate at 1–2 gpm/ft$^2$ (2.4–5 m/h) and have a smaller footprint than conventional flocculation and sedimentation; however, these have not been used in seawater desalination applications. |
| **Microsand-enhanced settling (MES)** | MES (Actiflo) can be used for clarification upstream of conventional filtration. MES uses microsand attached to the coagulated solids by a polymer to enhance the clarification process. | MES operates at 16–35 gpm/ft$^2$ (39–87 m/h), loading rates significantly higher than conventional sedimentation, plate settlers, and DAF, and has a very small footprint. MES has not yet been employed at full-scale SWRO facilities, and there is concern over the impact of the polymers on membrane performance. |

*Source:* Larry VandeVenter

at a reduced filter loading rate, followed by diatomaceous earth filtration as an additional barrier to prevent mussel related issues.

Recently installed SWRO plants have used a variety of different pretreatment approaches depending on the type of seawater intake and the feedwater quality. Table 3-3 summarizes some of these installations (IDA 2021). Pretreatment systems have included single- and two-stage filtration, sedimentation, DAF, UF, and diatomaceous earth filtration.

In facilities employing UF or MF, an additional screening step is required to prevent damage to the MF/UF membranes. While 500-μ strainers are commonly used in freshwater treatment facilities employing MF and UF, the prevalence of barnacles in seawater, which can be 130–150 μ during their embryonic phase, necessitates the use of tighter microstrainers with mesh sizes of 80–120 μ. Microstrainers that have been employed or successfully piloted at SWRO facilities include plastic rotating disc strainers, rotating drum screens, and automatic backwashing strainers. In cases where additional pretreatment equipment, such as DAF, has been provided before the MF/UF membranes, the screening requirements may be relaxed.

**Table 3-3 Partial list of pretreatment installations in seawater reverse osmosis (SWRO) plants since 2005**

| Installed Date | Location | Plant Name | Capacity mgd | Capacity ML/d | Intake Type | Pretreatment |
|---|---|---|---|---|---|---|
| 2022 | UAE | Umm Al Quwain | 180 | 683 | Open sea | DAF, gravity filters |
| 2022 | Saudi Arabia | Rabigh | 159 | 600 | Open sea | Dual media filters, UF |
| 2021 | Saudi Arabia | Shuqaiq 3 | 119 | 450 | Open sea | Dual media filters |
| 2021 | Singapore | Jurong Island | 36 | 137 | Offshore, open sea | DAF, UF |
| 2020 | Oman | Salalah | 30 | 114 | Open sea | DAF, dual media filters |
| 2019 | Saudi Arabia | Shuaiba 3 | 66 | 250 | Open sea | Dual media filters |
| 2018 | Oman | Barka | 74 | 281 | Open sea | Multimedia filters, MF |
| 2017 | Chile | Escondida | 76 | 288 | Offshore, open sea | Dual media pressure filters |
| 2015 | Israel | Ashdod | 72 | 274 | Offshore, open sea | UF |
| 2015 | California | Carlsbad | 50 | 189 | Shoreline, open sea | Dual media pressure filters |
| 2014 | Algeria | Magtaa | 132 | 500 | Offshore, open sea | UF |
| 2014 | Saudi Arabia | Ras Al Khair | 81 | 307 | Open sea | DAF, dual media filters |
| 2013 | Israel | Sorek | 165 | 624 | Offshore, open sea | Dual media gravity filters |
| 2013 | Singapore | Tuas 2 | 85 | 320 | Offshore, open sea | DAF, UF |
| 2013 | Saudi Arabia | Jubail 2 | 16 | 59 | Open sea | UF |
| 2012 | Australia | Victoria | 108 | 410 | Offshore, open sea | Pressure dual media filters |
| 2012 | Australia | Adelaide | 72 | 274 | Offshore, open sea | UF |
| 2012 | Trinidad | Point Lisas | 48 | 182 | Shoreline, open sea | Gravity filters, MF |
| 2011 | Australia | Perth II | 72 | 274 | Offshore, open sea | UF |
| 2011 | UAE | Fujairah II | 136 | 35.9 | Offshore, open sea | DAF, single-stage dual media gravity filters |
| 2010 | Australia | Sydney | 66 | 250 | Offshore, open sea | Dual media gravity filters |
| 2010 | Israel | Hadera | 72 | 274 | Offshore, open sea | Dual media filters |
| 2009 | Spain | Barcelona | 53 | 200 | Offshore, open sea | Pressure filters |
| 2009 | Australia | Gold Coast | 133 | 35.1 | Offshore, open sea | Single-stage dual media gravity filters |
| 2008 | Algeria | Skikda | 100 | 26.4 | Open sea | Two-stage dual media pressure filters |
| 2008 | UAE | Sharjah | 22.7 | 6.0 | Offshore, open sea | Two-stage dual media pressure filters |
| 2008 | Algeria | Hamma | 200 | 52.8 | Offshore, open sea | Lamella clarifiers, single-stage dual media gravity filters |
| 2008 | Aruba | Balashi | 8 | 2.1 | Shoreline, open sea | Pressurized single-stage sand filters |
| 2007 | Florida | Tampa Bay | 109 | 28.7 | Shoreline, open sea | Flocculation, single-stage upflow sand filters, diatomaceous earth filters |

*(continued)*

**Table 3-3 Partial list of pretreatment installations in seawater reverse osmosis (SWRO) plants since 2005 (*Continued*)**

| Installed Date | Location | Plant Name | Capacity mgd | Capacity ML/d | Intake Type | Pretreatment |
|---|---|---|---|---|---|---|
| 2007 | Israel | Palmachim | 110 | 29.0 | Offshore, open sea | Single-stage dual media gravity filters |
| 2007 | Australia | Perth | 144 | 37.9 | Offshore, open sea | Single-stage dual media pressure filters |
| 2007 | Singapore | Power Seraya | 10 | 2.6 | Shoreline, open sea | Two-stage dual media pressure filters |
| 2006 | Bahamas | Blue Hills | 27 | 7.2 | Seawater wells | Cartridge filtration |
| 2006 | Saudi Arabia | Kindasa | 25.5 | 6.7 | Shoreline, open sea | Single-stage dual media filters and UF |
| 2006 | China | Yu-Han | 70 | 18.5 | Open sea | UF |
| 2005 | Israel | Ashkelon | 326 | 86.1 | Offshore, open sea | Single-stage dual media gravity filters |
| 2005 | Japan | Fukuoka | 50 | 13.2 | Offshore infiltration gallery | UF |
| 2005 | Singapore | Tuas | 136 | 36.0 | Offshore, open sea | DAF, single-stage dual media filtration |

DAF — dissolved air flotation; MF — microfiltration; UF — ultrafiltration

## 3.2 SWRO DESIGN PARAMETERS

SWRO systems typically consist of feedwater pumps, RO membrane elements installed in pressure vessels, supporting framework and racks for the pressure vessels, valves and piping, instrumentation and controls, and sample panels. Either single- or two-stage RO can be used depending on finished water quality goals. Energy recovery devices (ERDs) have become industry standard in seawater applications to reduce kinetic energy requirements and therefore to lower operational costs. Cartridge filters are required as a safety barrier to protect RO membranes from particulate/colloidal matter that may be present in the filtered feed water. The main parameters for SWRO design include flux, recovery, salt rejection, and operating temperature. Each of these is discussed briefly in the following sections. In addition to these parameters, source water total dissolved solids (TDS) will have a considerable impact on operating pressures and facility cost. Where source waters are available with salinity lower than the open ocean, these waters are often preferred.

### 3.2.1 RO Flux

The membrane flux normalizes water production rate (permeate) per membrane surface area and can be expressed as gallons per day per square foot (gpd/ft$^2$) of membrane area, gallons per square foot per day (gfd), or in metric units as liters per meter squared per hour (LMH). Flux has a major impact on both the capital and operating costs of the facility, where a lower flux will require more membrane area, higher capital costs, and a larger plant footprint in comparison to higher fluxes. However, higher fluxes require higher feed pressures that result in increased energy use and higher operating costs. Use of higher fluxes may also increase fouling, causing further increase in operating costs. The Affordable Desalination Collaboration that conducted a flux study program in southern California in the early 2000s investigated operations in the range of 6–9 gfd (10–15 LMH) and concluded that this range may result in lower overall life-cycle costs for desalination. Most existing SWRO facilities operate at fluxes of approximately 8–12 gfd (14–20 LMH). An average flux of 10 gfd (17 LMH) is a default number for budgeting purposes when conditions are unknown.

An additional result of operating at higher fluxes is increased salt rejection. This phenomenon may seem counterintuitive, but it is the result of a higher ratio of product water diluting the relatively constant passage of salt across the membranes. Increasing the flux will increase this dilution factor, improving the quality of the product water. Ultimate determination of the most appropriate flux for a specific SWRO facility involves balancing capital costs, operating costs, and desired product water quality.

### 3.2.2 Recovery

Product water recovery is the ratio of the product water flow (permeate) rate to the feedwater flow rate, representing the overall production efficiency of the desalination system. Higher recoveries result in smaller intakes, smaller pretreatment facilities, and lower concentrate flows but also produce more concentrated residual streams and higher salinity product water. For example, 50-percent recovery will concentrate the TDS levels in the source water by twofold, while 75-percent recovery will increase TDS levels fourfold. Elevated concentrations of sparingly soluble salts in the concentrate cause higher scaling potential, and elevated overall salinity concentrations increase the osmotic pressure of the feed water requiring higher operating pressures.

Brackish water and reclaimed water treatment facilities typically operate at recoveries of 70–80 percent (sometimes higher) because these water sources have lower TDS levels than seawater; recovery in these cases is generally limited by the scaling potential of sparingly soluble salts that may be present. Higher TDS levels in seawater result in higher feed pressure requirements, and recoveries in SWRO facilities typically are 40–60 percent, which in part reflects the balance between operational costs (feed pump energy demands and pretreatment requirements) and capital costs (overall plant footprint and membrane system requirements). In addition, the residuals volume (membrane concentrate) will decrease as recovery increases, so the costs of residuals management must also be considered.

Figure 3-1 provides the power consumption of SWRO systems as a function of system recovery for three different feedwater temperatures (adapted from Wilf and Klinko 2001). While power consumption will vary with membrane type, flux, and energy recovery measures employed, the general trends shown in Figure 3-1 should apply.

In addition to the impact on energy use, a higher recovery will increase the salt concentrations in the permeate, resulting in a poorer product water quality.

### 3.2.3 Feedwater Temperature

For SWRO systems, an increase in feedwater temperature results in lower feed pressure requirements and therefore lower energy use when flux and recovery are maintained at constant conditions. Figure 3-2 shows this effect using manufacturer software projections of the feed pressure as a function of temperature for two different flux rates and two different SWRO membrane element types (Hydranautics SWC5 and SWC4+ elements, 50-percent system recovery, and typical seawater quality provided in Table 2-1). As this figure shows, an increase from 15 to 25°C (59 to 77°F) can decrease the projected feed pressure requirements by as much as 100 psi (690 kPa).

In addition to affecting feed pressure requirements, changes in the influent water temperature also affect salt passage. Higher temperatures increase the salt passage, resulting in a lower water quality (higher salt content). If boron and chloride levels are

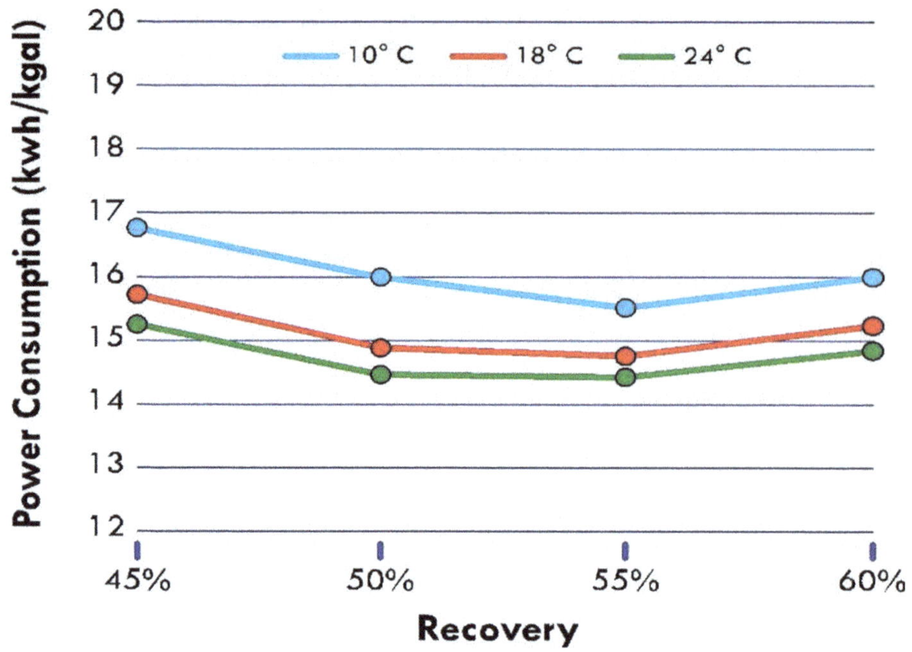


*Adapted from* Wilf and Klinko 2001

**Figure 3-1 Projected impact of recovery on power consumption for seawater reverse osmosis (SWRO)**

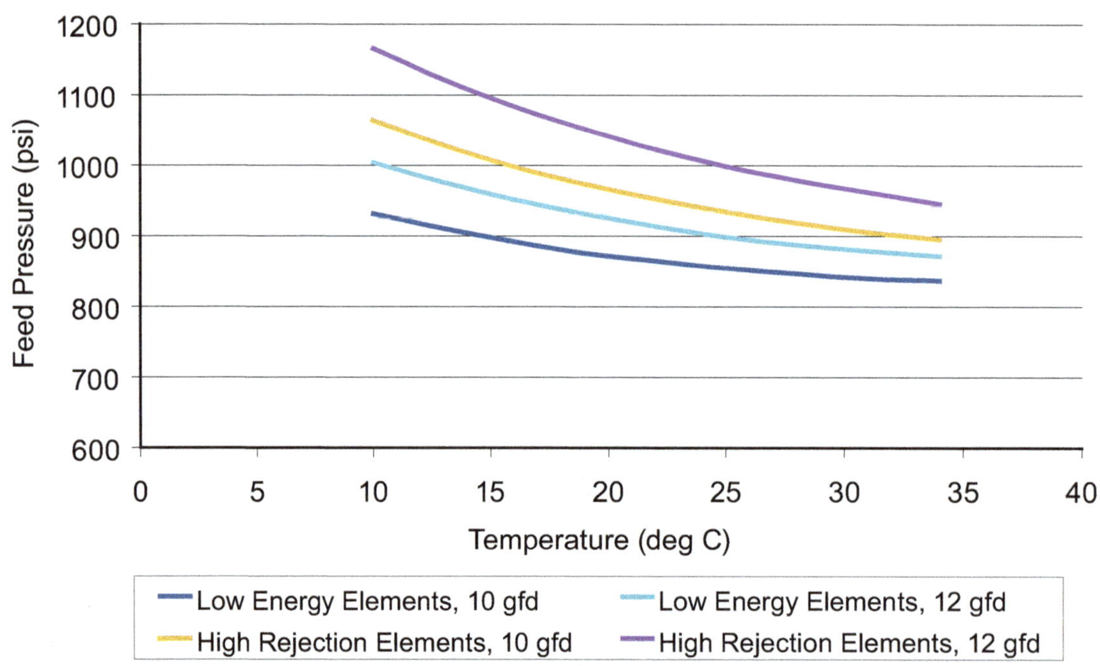


*Source:* Ken Klinko and Greg Wetterau using Hydranautics IMSDesign (2009)

**Figure 3-2 Projected seawater reverse osmosis (SWRO) feed pressure requirements as a function of influent water temperature for different flux rates and element types**

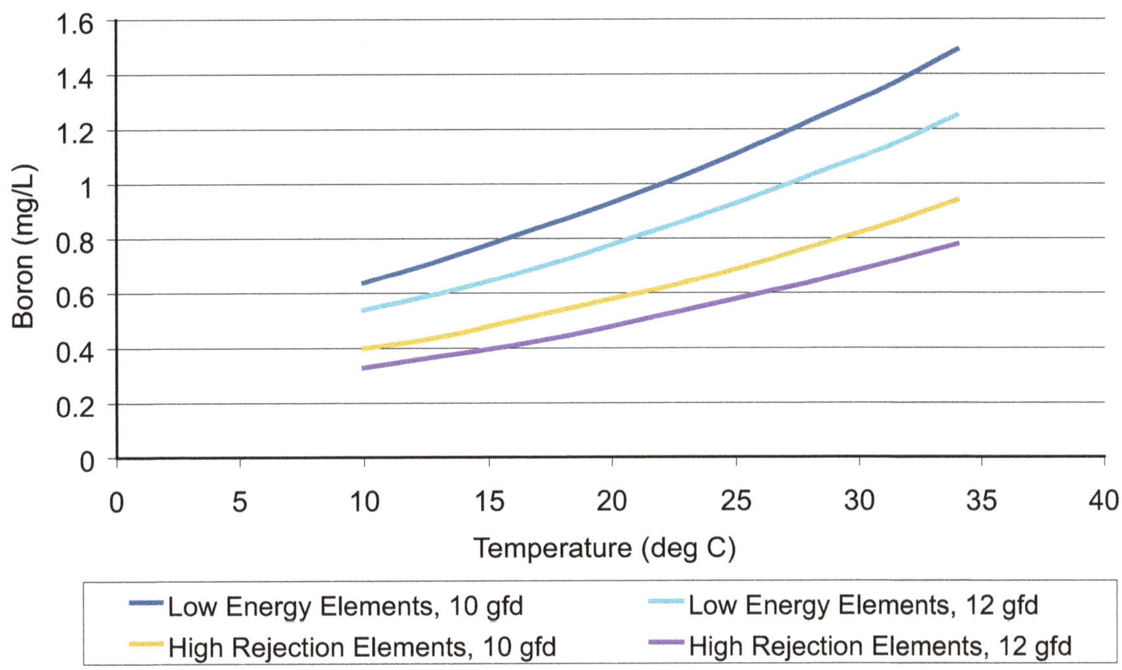


*Source:* Ken Klinko and Greg Wetterau using Hydranautics IMSDesign (2009)

**Figure 3-3 Projected impact of temperature on seawater reverse osmosis (SWRO) permeate boron**

constraining parameters in product water quality, this increased salt passage can result in a water quality that may not meet the treatment goals, depending on the flux, recovery, and membrane type used.

The relationship between feedwater temperature and permeate boron concentration is presented in Figure 3-3; this figure was generated using manufacturer software projections for two membrane types and two flux rates (50-percent recovery with typical seawater quality identified in Chapter 2). This figure indicates that a boron goal of 1.0 mg/L can be achieved at temperatures below 22°C (71.6°F) for either membrane type at fluxes greater than 10 gfd (17 LMH). Using higher boron rejection membranes, this goal can be met at temperatures as high as 34°C (93.2°F).

A lower boron goal of 0.5 mg/L can also be met using the high boron rejection membranes but only at temperatures up to 15°C (59°F) for a flux rate of 10 gfd (17 LMH) and for temperatures up to 20°C (68°F) for a flux rate of 12 gfd (20 LMH). The newer criteria for boron in potable water set by World Health Organization have been set at 2.4 mg/L. However, boron concentrations may be limited by the potential agricultural use of the water. Some countries and several US states are considering or have implemented boron action levels of 1.0 mg/L for this reason.

### 3.2.4 Feed Pressure

The standard seawater desalination equipment and membranes are designed for a maximum operational pressure of 1,200 psi (8,000 kPa). This limit is set by the mechanical strength of the equipment, the potential for membrane compaction, and economics. However, several efforts were conducted to consider higher operational pressure to achieve more economical designs. One of these concepts is ultrahigh pressure (UHP) desalination, introduced by Hydranautics (Oceanside, Calif.). The UHP designed for operational pressure 1,700 psi (12,000 kPa) can utilize a two-stage RO system similar to brackish water desalination and a turbocharger.

## 3.3 DISINFECTION

In the United States, disinfection requirements for drinking water supplies are established by individual states based on two guidelines established in the US Environmental Protection Agency Surface Water Treatment Rule (SWTR): Long Term 2 Enhanced SWTR (LT2) and the Stage 2 Disinfectants and Disinfection Byproducts Rule (Stage 2 D/DBPR). These water quality issues are more fully described in Chapter 2.

Table 3-4 includes a summary of the disinfection log reduction credits given to typical treatment processes. It should be noted that disinfection credits will vary based on dose, contact time, water quality, and other operating conditions, while credits given for membrane processes will vary by state. It is assumed that seawater would typically be placed in bin 1 based on LT2 criteria, which will not require additional treatment for *Cryptosporidium* removal or inactivation; however, it is important to consider the possibility of seawater contamination by wastewater treatment plant discharges if the plant is located close to open seawater intakes.

The Department of the Drinking Water (DDW) of the California Water Boards (formerly the California Department of Public Health) has granted RO membranes 2-log removal credits for viruses, *Giardia*, and *Cryptosporidium* based on the continuous monitoring of 2-log TDS reduction, using online conductivity analyzers. Other surrogate measurements are being evaluated to possibly give higher removal credits; however, most states have yet to adopt formal policies on pathogen removal credits for RO membranes. Selection of the treatment process should therefore consider the requirements and specific credits given by the individual state in which the plant is operating.

Desalinated water can also impact disinfection residual management in the distribution system if chloramines are used. The primary concern is that bromide in the desalinated water will impact the initial chlorine demand and cause rapid decreases in disinfectant residuals once the water is blended with other sources in the distribution system, which may cause temporary noncompliance with regulations.

As discussed in Chapter 2, blending of desalinated water containing bromide with water from distribution systems containing treated surface water can have potential impacts on DBP formation. Desalinated seawater is low in TOC, which can serve as a low precursor to DBP formation; however, when the desalinated seawater is blended with higher TOC surface water supplies, the DBP formation in the blend can be greater than in either of the waters individually. Blending studies are therefore recommended where DBP formation is a concern.

**Table 3-4 Log removal credits for potential treatment processes**

| | Log Removal Credits Required* | | Log Removal and/or Inactivation Credits Allocated for Typical Treatment Processes | | | | | |
|---|---|---|---|---|---|---|---|---|
| Pathogen | Minimum | Maximum | Conventional Treatment | MF | UF | RO§ | UV | Chlorination |
| Virus | 4 | 6.0† | 2 | 0–1 | 0–1 | 0–2 | 0 | 2–4 |
| *Giardia* | 3 | 5.0† | 2.5 | 4 | 4 | 0–2 | 3–4 | 0–0.5 |
| *Cryptosporidium* | 2 | 5.5‡ | 3 | 4 | 4 | 0–2 | 3–4 | 0 |

*Source:* Greg Wetterau

MF — microfiltration; RO — reverse osmosis; UF — ultrafiltration ; UV — ultraviolet light

* States may require a minimum 0.5 log of *Giardia* or 2-log of virus disinfection beyond filtration to provide a multibarrier treatment approach.

† For California only, based on Division of Drinking Water guidelines for impaired water sources.

‡ Based on Long Term 2 Enhanced Surface Water Treatment Rule for bin 4.

§ Removal credits not currently granted for RO in Texas, while California has granted up to 2-log based on total dissolved solids reduction. Removal credits for RO may differ as more states adopt policies on seawater desalination.

## 3.4 POSTTREATMENT

Because RO treatment removes most of the hardness and alkalinity of the source seawater, RO permeate must be remineralized to prevent corrosion in distribution system piping and to produce a finished water that is aesthetically acceptable to the customers. It is typically best to match existing distribution system water quality to the maximum extent possible to avoid customer complaints or problems with pipeline corrosion or release of scale in the distribution system. Such concerns are common to all RO designs; however, unlike brackish RO systems, where a percentage of raw water can be bypassed around the RO for product water stabilization, SWRO facilities must make use of an external source of hardness and alkalinity.

In SWRO facilities, hardness is most commonly added with either calcium oxide (quicklime), calcium hydroxide (hydrated lime), or calcium carbonate (limestone or calcite). Quicklime is typically the least costly alternative, but it requires slakers that are operationally intensive and sometimes seen as undesirable by plant operators. In addition, the use of quicklime or hydrated lime at the tail end of the treatment process will likely require lime saturators to avoid problems with turbidity from inert materials and incomplete dissolution of the lime. These saturators will add to the cost and operating complexity of the lime feed system but may be necessary to produce the high-quality and low-turbidity product water.

Calcite or limestone contactors are the simplest approach for remineralization from an operations perspective, but these require large footprints for calcium saturation to occur and can be a costly approach for large treatment facilities. Calcite contactors are common for small desalination facilities, such as the 0.3 mgd (1.1 ML/d) facility in Sand City, Calif.; however, they were also employed in the 52.8 mgd (200 ML/d) Barcelona facility in Spain and were utilized for the 50 mgd (189 ML/d) desalination plant in Carlsbad, Calif., which is the largest operational seawater desalination plant in the United States currently. The best approach for remineralization should consider both economic factors and operations preferences for the utility.

Carbon dioxide is frequently added to provide alkalinity, and sodium hydroxide is often added to adjust the pH to match the existing pH within the distribution system. Other alternatives for product water stabilization include the use of calcium chloride with sodium hydroxide, use of orthophosphate or polyphosphate, or blending with a readily available source of hard water like local surface water or groundwater.

## 3.5 ENERGY RECOVERY

Energy represents the single greatest major cost for operating SWRO facilities due to the feed pressure requirements for RO membranes. To reduce the volume of feedwater pumped, the most common approach is to increase system hydraulic recovery. Higher hydraulic recovery during the desalination process leads to more concentrated brine, which in turn has a higher osmotic pressure leading to an even higher feed pressure. Much of this pressure still remains in the concentrate waste stream, with the inherent energy sometimes wasted, burned over an orifice plate or throttling valve downstream of the membranes. The industry standard is to include an energy recovery system to recapture the kinetic energy remaining in the concentrate stream, thus significantly reducing the energy input required for the feedwater.

### 3.5.1 ERDs

There are now a variety of commercially available ERDs that allow energy recovery from the brine stream to reduce electrical pumping costs. The primary types of ERD can be divided into two major categories: centrifugal devices and positive displacement devices. Centrifugal devices include the Francis turbine, the Pelton wheel turbine, and the hydraulic turbocharger (HTC); positive displacement devices include the work exchanger and pressure exchanger.

Each type of ERD is described in the following sections.

**Francis turbine.** In a Francis turbine, water enters the turbine runner with a radial velocity component and discharges with an axial velocity component, like a reverse running pump. Francis turbines are distinguished by having a band that surrounds the peripheral end of the blades (also known as buckets), providing a boundary for the water passage and structural rigidity to the runner. Francis turbines are directly coupled to the feed pump and must be designed for specific operating conditions. The result is that changes in flow and pressure must be bypassed around the unit, lowering recovery efficiency.

**Pelton impulse turbines.** The Pelton wheel turbine operates by converting the velocity energy from a brine stream into kinetic energy. Nozzles aim the pressurized concentrate stream toward the Pelton wheel, which rotates a turbine, creating electrical energy to assist the electric motor in driving the high-pressure feed pumps. Up to 80 percent of the concentrate energy can be recovered using this device. The initial capital cost is relatively high because it must be incorporated into the feed pump and must employ high alloy stainless steels to be compatible with seawater concentrate. Figure 3-4 shows a photo of Pelton wheel generators installed at an SWRO facility.

**Figure 3-4 Pelton wheel generator at a seawater reverse osmosis (SWRO) facility**

**HTCs.** An HTC is an integral, centrifugal feed pump with an energy-recovery turbine coupled by a shaft. The HTC receives pressure from the RO concentrate and returns it to the RO system as a pressure boost to the feed stream resulting in reduced feed pressure required from the RO pressure pump. The turbocharger has no brine to feed mixing when the energy recovery process occurs. Moreover, no auxiliary equipment like a booster pump, variable frequency drive, high-pressure flow meter, or manifold is required to attain its purpose of recovering energy from the brine. The latest HTCs report peak efficiencies of 86 percent, with efficiency dropping as flow and pressure change from the primary design point by 2.5 percent. Turbochargers have been popular where footprint, plant availability, and/or capital expenditure were the main drivers of the project; in addition, turbochargers have been successfully used as efficient interstage boosters in brackish, seawater, and zero liquid discharge applications. Figures 3-5 and 3-6 illustrate how an HTC works in an RO system, and Figure 3-7 shows an HTC.

**Isobaric pressure exchangers.** A PX™ pressure exchanger device transfers brine pressure energy directly to a portion of the incoming feedwater. A booster pump then makes up the difference required to achieve the needed feed pressure. This stream then joins the portion of the feed from the high-pressure feed pumps that has not passed through the device. The PX™ pressure exchanger device has a single moving part, a shaftless ceramic rotor that is suspended within a sleeve. PX™ devices can recover up to 98 percent of the energy remaining in the concentrate. Their high efficiency has contributed to their increased use at many recently built SWRO facilities. Figure 3-8 shows a flow diagram of a PX™ device from Energy Recovery, Inc., installed on an RO skid. Figure 3-9 depicts an image of a typical PX™ ERD installation.

**Dual work exchangers (DWEER).** In a work exchanger configuration, the high-pressure brine is directed to a work exchanger vessel filled with seawater to pressurize the influent seawater. A small recirculating pump boosts the pressure of the seawater exiting

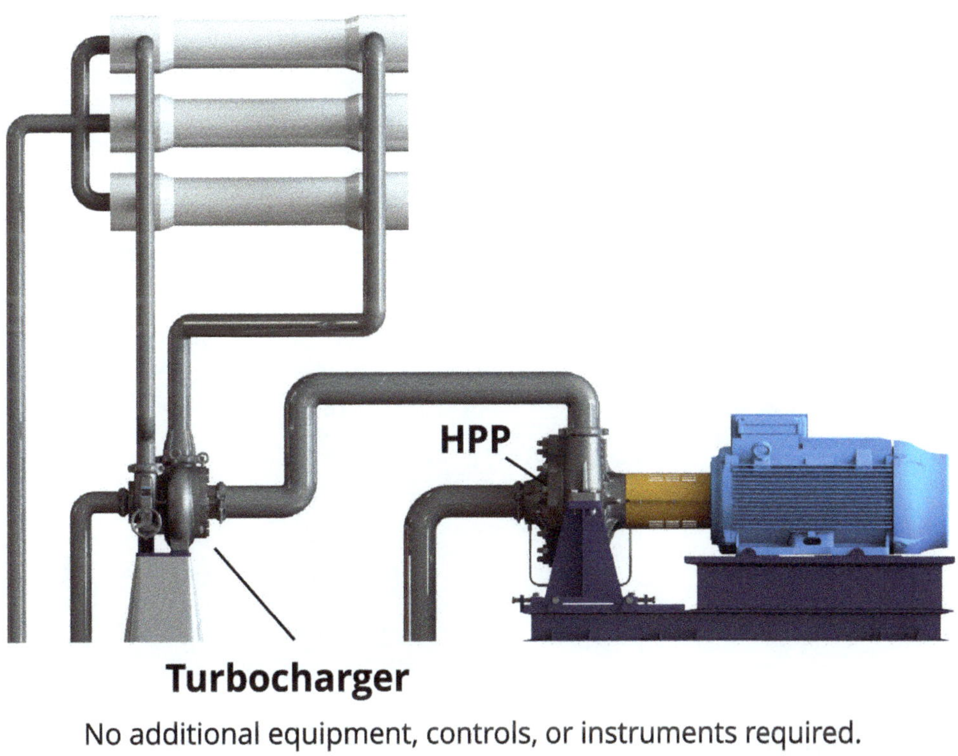


*Source:* FEDCO-USA

**Figure 3-5 Hydraulic turbocharger (HTC) in a single-stage reverse osmosis (RO) system.**

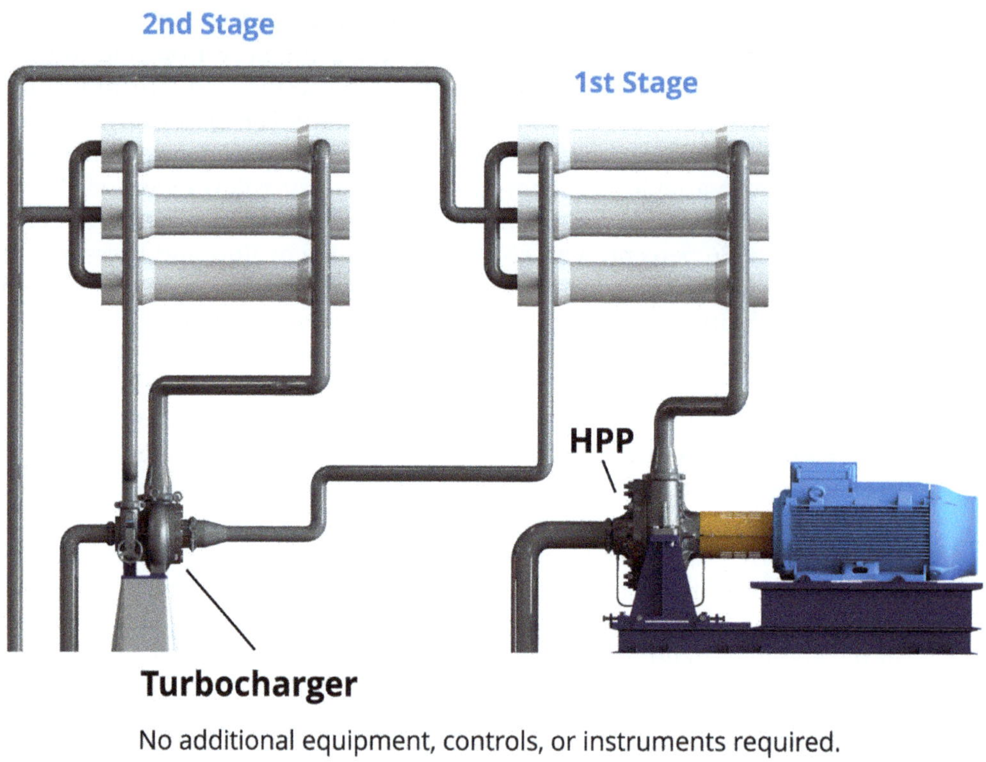


*Source:* FEDCO-USA

**Figure 3-6 Hydraulic turbocharger (HTC) in an interstage configuration for brackish water reverse osmosis (BWRO), seawater reverse osmosis (SWRO), and zero liquid discharge applications**

*Source:* FEDCO-USA

**Figure 3-7 One hydraulic turbocharger (HTC) for a 5.7-mgd train at Ras Al Khair (360,000 m$^3$/d [97 mgd]), Saudi Arabia, commissioned in 2015. Only one ERD per train, no auxiliary equipment required. Approximate volumetric footprint is 1 m$^3$ (35 ft$^3$).**

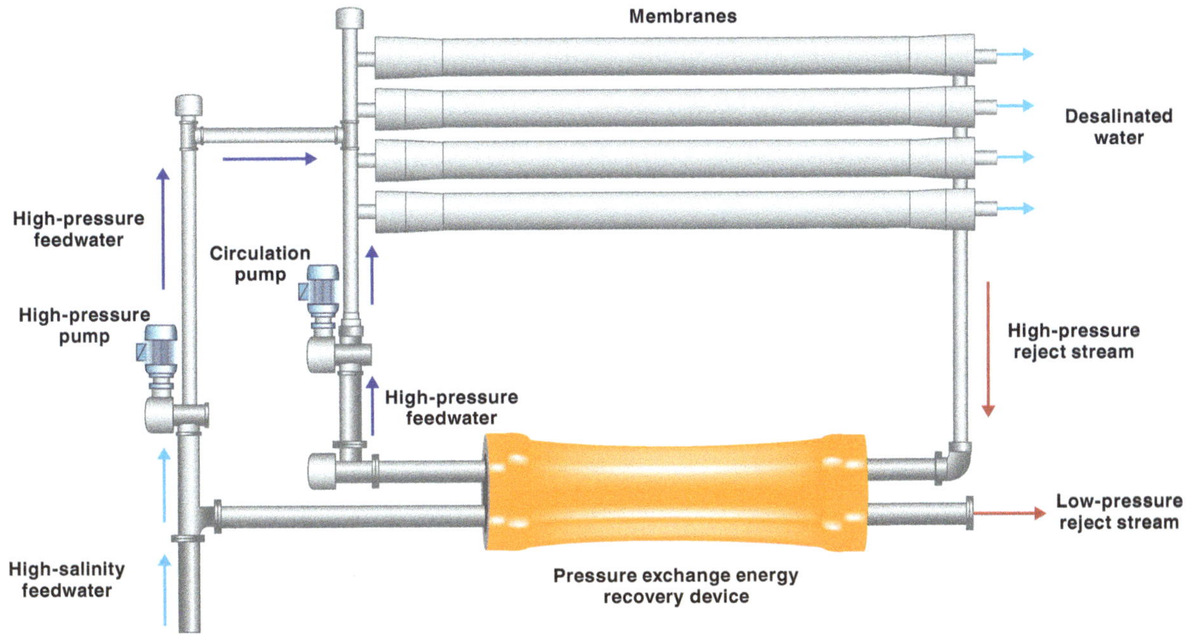


**Figure 3-8 Isobaric energy recovery device (ERD) flow diagram**

*Source:* Greg Wetterau

**Figure 3-9 Isobaric PX™ pressure exchanger device (ERD) installation in Sand City, Calif.**

the work exchanger vessel to equal the main feed pump pressure and joins the flow to the membranes. Efficiencies of the work exchanger piston system can be up to 97 percent, similar to the pressure exchanger. These devices are more efficient than centrifugal designs, which rely on shaft conversion of power. Figure 3-10 shows a typical DWEER flow diagram, while Figure 3-11 depicts a DWEER installation in the Bahamas.

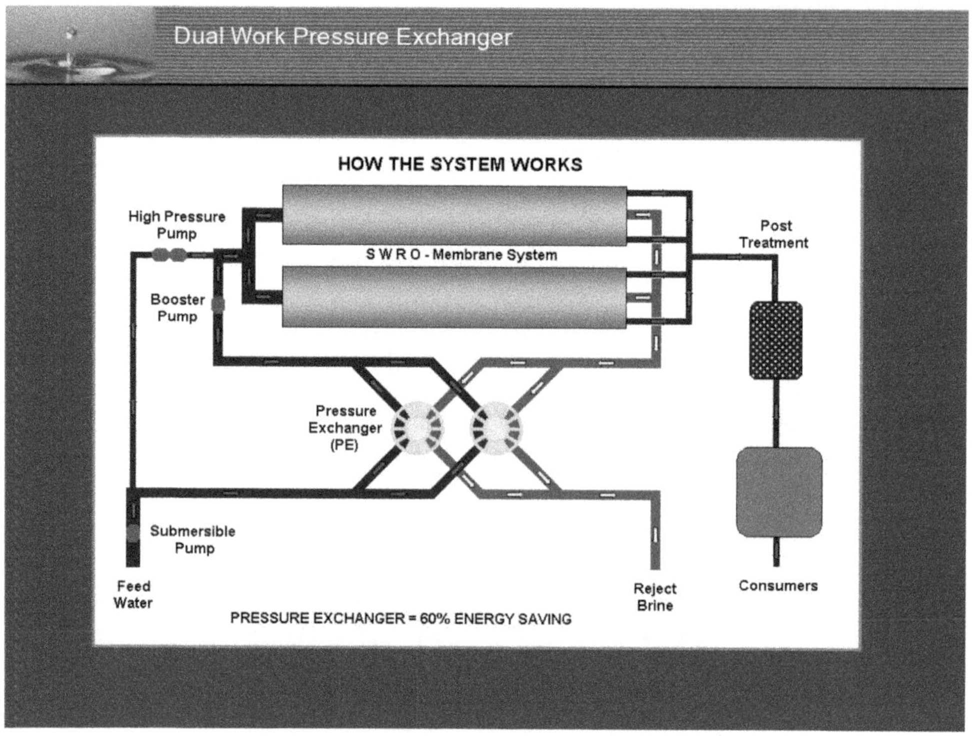


*Source*: Val S. Frankel

**Figure 3-10 Dual work pressure exchanger (DWEER) flow diagram**

*Source*: Srinivas Veerapaneni

**Figure 3-11 Installation of flowserve dual work exchangers (DWEER) energy recovery device (ERD)**

ERDs can be installed separately on each RO train, or a multitrain concept can be applied at large SWRO plants where both RO feed pumps and ERDs are arranged to serve multiple RO trains. The flow schematic of the 88 mgd (333 ML/d) Ashkelon SWRO plant illustrates a multitrain approach shown in Figure 3-11. The three-center design layout is detailed in Figure 3-12.

### 3.5.2 Summary

Introduction of ERDs and other developments in desalination have allowed reduced energy demand for SWRO desalination by almost three times from 27 kW•h/kgal (7.1 kWh/$m^3$) in 1980 to as low as 10 kW•h/kgal (2.6 kW•h/$m^3$) currently. ERDs themselves allow the recovery of nearly half the energy used in feed water pumping. The technologies are improving, and this tendency will continue as a result of continued research and different initiatives around the globe.

Each commercially utilized ERD technology has certain pros and cons, some of which are summarized in Table 3-5.

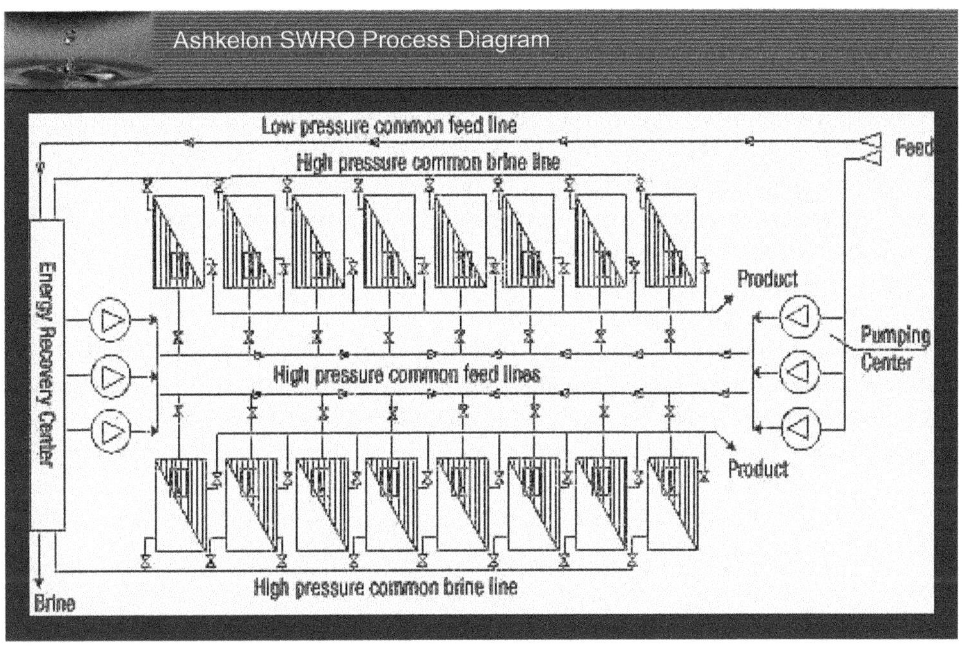


*Source:* IDE Technologies Ltd.

**Figure 3-12 Three-center design layout**

**Table 3-5 Energy recovery devices (ERDs): pros and cons**

| ERD Type | ERD Category | Maximum Efficiency (%) | Concentrate Mixing with Feedwater | Concentrate Repumping Required | Sidestream Booster Pump Required |
|---|---|---|---|---|---|
| Reverse running pumps | Centrifugal devices | 75 | No | No | No |
| Pelton impulse turbines | Centrifugal devices | 80 | No | Yes | No |
| Hydraulic turbochargers | Centrifugal devices | 65 | No | No | No |
| Pressure exchangers | Positive displacement devices | 98 | Yes | No | Yes |
| Dual work exchangers | Positive displacement devices | 97 | Yes | No | Yes |

*Source:* Val S. Frankel

## 3.6 CORROSION AND MATERIALS OF CONSTRUCTION

In the design of desalination facilities, potential corrosion issues require serious attention. Within an SWRO plant, a number of potential corrosion problems may occur as a result of the aggressive environment of seawater and concentrate in which the equipment must function. Nonmetallic materials are often used, wherever possible, to reduce corrosion impacts; however, the high pressures seen in SWRO facilities require that a major portion of the piping, valves, and fittings be metallic in construction. A number of factors can influence corrosion rates in metallic materials, including the chemical composition of the material, heat treatment utilized during manufacturing, fabrication techniques, surface condition, and design. A number of environmental factors can also influence corrosion, including solution composition, pH, temperature, oxygen content, flow velocity, deposited solids, and microbial growth. As a result, the design of SWRO facilities should carefully consider and evaluate the materials of construction.

### 3.6.1 Corrosion Types and Mitigation

Corrosion types of concern in SWRO plants include crevice, galvanic, pitting, microbial, under-deposit, and stress corrosion. Each type of corrosion and associated mitigation approaches are discussed in the following sections.

**Crevice corrosion.** Crevice corrosion occurs when local differences in oxygen concentration occur at a metal surface, typically associated with small pockets of stagnant water in voids. These voids can exist as deposits on the metal surface like foulants or welding slag, gaskets, lap joints, or under bolt or rivet heads.

Crevice corrosion can be mitigated by minimizing crevices or other conditions that can lead to seawater stagnation, using materials resistant to crevice corrosion (Figure 3-13), and using properly constructed butt-welds as opposed to socket welds.

**Galvanic corrosion**. A number of different metals are typically used to construct a desalination system. Galvanic corrosion is an electrochemical process in which one metal corrodes preferentially when in electrical contact with a different type of metal while both metals are immersed in an electrolyte. When two dissimilar metals are physically connected, the metal with the more negative potential generally has increased corrosion compared to its corrosion rates when not coupled with another metal. Table 3-6 presents a galvanic series list for common metallic materials that may be used in SWRO facilities.

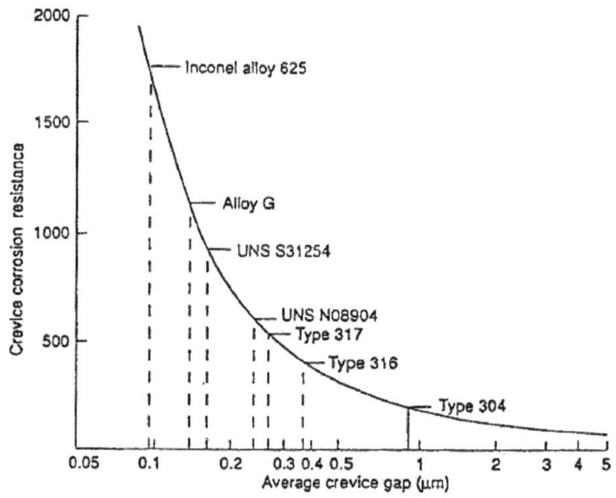


*Source:* Oldfield and Todd 1991

**Figure 3-13 Resistance to crevice corrosion**

**Table 3-6 Galvanic series for alloys in flowing seawater at 4 m/s and 24°C**

| Material | Steady State Electrode Potential (V) v SCE |
|---|---|
| Graphite | +0.25 |
| Platinum | +0.15 |
| AL6XN (super austenitic stainless steel) | 0.00 |
| Zeron 100 (super duplex stainless steel) | –0.01 |
| AISI type 316 stainless steel (passive) | –0.05 |
| AISI type 304 stainless steel (passive) | –0.08 |
| Hastelloy alloy C | –0.08 |
| Titanium | –0.10 |
| AISI type 410 stainless steel (passive) | –0.15 |
| AISI type 316 stainless steel (active) | –0.18 |
| Nickel | –0.20 |
| AISI type 430 stainless steel (passive) | –0.22 |
| Copper alloy (70Xu-30Ni) | –0.25 |
| Copper alloy (90Cu-10Ni) | –0.28 |
| Copper | –0.36 |
| AISI type 304 stainless steel (active) | –0.53 |
| Carbon steel | –0.61 |
| Cast iron | –0.61 |
| Zinc | –1.03 |

*Adapted from* Cramer and Covino, Jr. 2005.

Galvanic corrosion can be mitigated by using materials adjacent to each other in the galvanic series, ensuring that the key component is of a more noble material, by ensuring that the less noble material is present in a much larger area than the more noble material, and by using insulation to decouple dissimilar metals (e.g., nonconductive sleeves and gaskets). Simply coating or painting metal surfaces to mitigate galvanic corrosion is of limited value and should be avoided.

**Pitting corrosion.** Pitting corrosion is a form of extremely localized corrosion that leads to the creation of small holes in the metal. The corrosion penetrates the metal with limited diffusion of ions, further pronouncing the localized lack of oxygen. Pits can range in size from those that are difficult to detect with the naked eye to those with a diameter and depth that can be measured in millimeters. Pitting occurs when the protective film of stainless steels breaks down in small, isolated spots. Pitting corrosion is typically associated with high chloride concentrations.

Pitting corrosion can be mitigated by using chromium-, molybdenum-, and nitrogen-based alloys. One percent molybdenum has approximately the same effect as 3 percent chromium. Resistance to pitting corrosion is rated by a pitting resistance equivalent number (PREN); the higher the PREN, the higher the resistance to pitting corrosion. Eq 3-1 shows how the PREN is calculated where element contents are given in weight percentage.

$$PREN = \%Cr + 3.3(\%Mo) + 16(\%N) \quad \text{(Eq 3-1)}$$

**Table 3-7 Pitting resistance equivalent number (PREN) values for common materials**

| Type | Material | Unified Numbering System Number | PREN |
|---|---|---|---|
| Austenitic | 304 L | S30403 | 18–20 |
| | 316 L | S31603 | 24–28 |
| | 317 L | S31703 | 29–31 |
| | 904 L | N08904 | 35–36 |
| Super austenitic | AL-6XN | N08367 | 47–48 |
| | 254 SMO | S31254 | 43–46 |
| Duplex | 2205 | S32205 | 35–37 |
| | 329 | S32900 | 34 |
| Super duplex | SAF 2507 | S32750 | 42 |
| | Zeron 100 | S32760 | ≥40 (42) |
| Austenitic | Alloy 20 (20Cb-3) | N08020 | 31 |
| | Hastelloy C-276 | N10276 | 68 |
| | Hastelloy C -22 | N06022 | 65 |
| | 90-10 Cu-Ni | C70600 | NA |

*Source:* Jacobs
NA — not applicable

Typically, austenitic and duplex stainless-steel piping used in desalination should have a PREN of 40 or more. Table 3-7 provides PREN values for materials commonly used in SWRO facilities.

**Microbial corrosion.** Microbially induced corrosion (MIC) occurs when microorganisms attach to surfaces like pipe walls and metabolize available nutrients using dissolved oxygen or other chemical compounds to generate corrosive byproducts (such as acids and sulfides). This results in corrosion of the metal, forming a pit or crevice beneath the biofilm area. Sulfate-reducing bacteria, iron and manganese bacteria, and sulfur-oxidizing bacteria have most often been associated with MIC of stainless steels.

Microbial corrosion can be mitigated by using shock chlorination and/or biocide addition, polymeric materials, or alloys high in molybdenum (>6 percent).

**Under-deposit corrosion.** Deposits, usually from corrosion of metal surfaces (such as iron oxide tubercles), provide electrochemical conditions that favor additional corrosion and buildup of deposits. In the worst cases, under-deposit corrosion causes deep pitting and pipe failure. The accumulated deposits also create favorable conditions for MIC.

Under-deposit corrosion is best mitigated by preventing surface corrosion in the first place. Once corrosion has occurred on the surface, under-deposit corrosion cannot really be prevented.

**Stress corrosion.** Stress corrosion and associated cracking are temperature dependent and can occur in materials exposed to both high tensile stresses and specific corrosion-inducing conditions. Stress corrosion of stainless steels in chloride containing water rarely occurs at temperatures below 70°C (158°F). This issue is typically associated with thermal desalination plants, as opposed to RO plants.

## 3.6.2 General Corrosion Mitigation Practices

The prevalence of corrosion is typically minimized by the following practices.

- Select proper material for the chemical conditions present.
- Use isolating sleeves, gaskets, etc., to decouple the metals from forming an active electrochemical cell.

- Use appropriate flushing to minimize stagnation of high-chloride water in piping. For intermittently used pipes, permeate flushes can be used.
- Low flow velocities and stagnant conditions increase the corrosion potential in austenitic stainless steels (but not super austenitic stainless steels) in high-chloride environments. Generally, piping velocities should not exceed 5 ft per s (fps) (1.5 m/s) upstream of pumps and not exceed 10 fps (3 m/s) in other parts of the system. High velocities may also result in scour and/or impingement. Design velocities for austenitic stainless-steel piping should be a minimum of 5 fps (1.5 m/s). When copper alloys are used, design per manufacturer recommendations for the specific product.
- As a cost-cutting factor, some vendors advocate the use of lined or painted components rather than more expensive components possessing appropriate corrosion resistance. Nonmetallic linings typically work well when whole but are prone to damage during construction or use. Once linings are breached, accelerated corrosion can occur, resulting in rapid failure, so the use of linings to mitigate corrosion should be considered carefully.
- Shock chlorination is frequently practiced to mitigate biofouling in SWRO systems but must be followed by complete dechlorination. Investigations have shown that chlorination increases the risks for localized corrosion in seawater (especially crevice corrosion). Continuously chlorinated seawater is considerably more aggressive than unchlorinated or intermittently chlorinated seawater. Increasing water temperature increases the corrosion potential in chlorinated seawater. The highest alloyed stainless steels are resistant to crevice corrosion, but lower-grade stainless steels are not. In contrast, the risk of galvanic corrosion decreases if the seawater is chlorinated.
- Many rotating machinery components and control system components require ventilation and cooling. Totally enclosed fan-cooled components, for instance, rely on airflow across the windings and use an integrated fan located on the end of the motor housing for cooling. Many electrical components that may accumulate heat like control panels, transformers, and variable frequency drives typically use fan-based cooling. Air in coastal regions is laden with seawater aerosols. These aerosols are electrolytes coat components being cooled, which may result in corrosion, short circuits, and electrical failures. External cooling, a dry air purge system, or another appropriate method is encouraged for cooling.
- Designers must account for seawater aerosols in the design of heating, ventilation, and air conditioning (HVAC) equipment. Dispersion of the aerosols through the HVAC system into the conditioned facilities should be discouraged. Accelerated corrosion of HVAC components exposed to seawater aerosols may happen.
- Most concrete is somewhat porous, allowing the migration/penetration of water (and seawater) into the concrete. Galvanic corrosion of rebar can result in the release of hydrogen gas and subsequent spalling of the concrete, substantially reducing the concrete's serviceable life. Mitigation can include some combination of the use of low permeability admixes with the use of coated rebar, which will ensure a minimum thickness of reinforced concrete over rebar, and the use of anodic protection of the rebar systems.
- While most valves can be constructed of corrosion-resistant materials, the use of dissimilar materials can result in corrosion. All of the valve trim must also be of appropriate materials. Air relief valves on seawater pipelines in desalination plants are often specified incorrectly, resulting in corrosion from trim materials.

- Appropriate selection of wetted materials of instrumentation is required. If not available, an isolating style diaphragm may be required to protect the instruments. Those most prone to failure are pressure instruments and orifice plate- or vortex-style flow meters.
- There has been an increasing trend to use plastic components as frequently as possible in desalination facilities. As a result, the industry is seeing an increased use of reinforced plastic for piping, pressure filter vessels, pumps, cartridge filter vessels, and other components. It is anticipated that this trend will accelerate.

## REFERENCES


Chave, K.E., and E. Suess. 1970. "Calcium Carbonate Saturation in Seawater: Effects of Dissolved Organic Matter." *Limnology and Oceanography*, 15:4:633–637.

Cheng, R., C. Andrews-Tate, T. Tseng, and K. Wattier. 2010. "Distribution of Desalinated Seawaters: Are Corrosion Indicators Sufficient?" *IDA Journal of Desalination and Water Reuse*, 2:1:24–31.

Cramer, S.D., and B.S. Covino, Jr., eds. 2005. *ASM Handbook Vol. 13B, Corrosion: Materials.* Materials Park, OH: ASM International.

Gisclair, M., C. Bartels, R. Chmielewski, and L. Tua Parra. 2022. *Established Practices, Innovative Design, and New Products: A New Generation of Ultra High Pressure Desalination* https://membranes.com/wp-content/uploads/Documents/Technical-Papers/Product%20line/RO/Established-Practices-Innovative-Design-and-New-Products_A-New-Generation-of-UHP-Desalination-1.pdf (accessed September 9, 2025).

IDA (International Desalination Association). 2021. *Desalination Yearbook.* https://idrawater.org/e-library/.

JICA (Japan International Cooperation Agency). 2022. Japan's Experiences on Water Supply Development. www.jica.go.jp/Resource/english/our_work/thematic_issues/water/c8h0vm0000ammj2q-att/activity_01.pdf (accessed September 9, 2025).

Oldfield, J.W., and B. Todd. 1991. *Reverse Osmosis—Which Stainless Steel to Use?* Toronto: Nickel Development Institute.

Voutchkov, N. 2008. "Pretreatment Technologies for Membrane Seawater Desalination." *Desalination*, 261:3:354–364.

Wilf, M., and C. Bartels. 2001. Optimization of seawater RO system design. *Desalination,* 173:1:1–12.

Chapter 4

# Environmental Impacts and Mitigation

*Second Edition Author: Kevin Thomas*
*First Edition Authors: Nikolay Voutchkov, Henry Hunt*

## 4.1 INTRODUCTION

The purpose of this chapter is to provide an overview of potential environmental impacts of seawater desalination facility construction and operation and to discuss potential strategies for avoiding, minimizing, or mitigating negative impacts.

The environmental impacts of seawater desalination facility operations have many similarities to those of conventional water treatment plants; however, seawater desalination facilities provide some unique environmental benefits as well as some risks of increased negative impacts. Like conventional water treatment facilities, desalination facilities have source water intake and waste stream discharge that may impact the aquatic environment in which they are located. In addition, desalination facilities and conventional water treatment plants may use many of the same chemicals for source water conditioning and therefore have similar waste streams, apart from salinity, associated with the disposal of the spent conditioning chemicals and the source water solids.

Some unique environmental benefits to seawater desalination facilities include

- an increased supply of high-quality freshwater from an essentially unlimited source, providing new water for both human consumption and environmental discharge;
- avoidance of increased withdrawals from sensitive or overtaxed freshwater ecosystems and aquifers; and
- potential for some positive environmental impacts associated with desalination brine outfalls, including increased fish abundance (Kelaher et al. 2022) and carbon sequestration opportunities (He and Tyka 2023).

**Table 4-1 Energy use for alternative freshwater production methods**

| Water Supply Alternative | Energy Use kW•h/1,000 gal (kW•h/m$^3$) |
|---|---|
| Conventional treatment of surface water | 0.8–1.5 (0.2–0.4) |
| Water reclamation | 1.9–3.8 (0.5–1.0) |
| Indirect potable reuse | 5.7–7.6 (1.5–2.0) |
| Direct potable reuse | 6.4–9.1 (1.7–2.4) |
| Brackish water desalination | 3.8–5.7 (1.0–1.5) |
| Seawater desalination by reverse osmosis | 9.5–15 (2.5–4.0) |

*Adapted from* WateReuse Association Desalination Committee 2012

Beyond the similarities to conventional water treatment plants and potential positive environmental impacts, seawater desalination facilities have several distinctive differences that could result in increased negative environmental impacts.

- They use approximately 1.5–2.5 times more source water to produce the same amount of freshwater. In fact, seawater desalination generally requires twice the ocean water volume as the desired product water volume depending on the reverse osmosis (RO) recovery rate, producing 1 gal (3.78 L) of product water and 1 gal of brine concentrate for every 2 gal (7.6 L) of ocean source water.
- They generate a discharge with elevated metals, salts, and other chemicals, which typically has 1.5–2 times higher salinity concentration than that of the source seawater.
- The ocean water desalination energy requirement is 8–10 times higher than conventional water supplies and approximately 33 percent more energy than some imported water supplies; therefore, it may have incrementally increased direct and indirect impacts on air quality, greenhouse gas emissions, and related issues.

These impacts vary widely depending on the ocean water salinity and comparative energy requirements of conventional or alternative water supplies (see Table 4-1).

The environmental impact of desalination facility operations should be assessed in the context of the environmental impacts of water supply alternatives that may be used instead of desalination. Desalination projects are typically driven by the limited availability of traditional lower-cost water supply sources, such as groundwater or fresh surface water (rivers, lakes, etc.). These traditional water supply sources may also result in environmental impacts from continued reliance on limited water supplies. For example, over-pumping of freshwater coastal aquifers may result in a significant increase in the salinity of the groundwater (known as seawater intrusion) and cause damage to these aquifers. In some arid areas, transfers of freshwater from a traditional water supply source like a river, river delta, or lake have impacted the ecosystem balance in this freshwater source to an extent that the long-term continuation of this water supply practice may result in significant and irreversible damage to the ecosystem of the traditional freshwater supply source and the wetlands created at the river mouth. In such cases, the environmental impact of the construction and operation of a new seawater desalination project should be weighed against the potentially environmentally damaging consequences from the continuation or expansion of the existing freshwater supply practices.

In addition, the impacts of a seawater desalination facility should be considered against the impacts of water reuse alternatives, both potable and nonpotable. Waste streams generated from desalination facilities are similar to those generated by conventional water treatment plants and water reuse facilities but with an additional concentrate stream. Water reclamation plants also generate waste streams that contain some of the same chemicals used for desalination and may also have elevated content of man-made waste substances that may impact the marine environment. Waste discharges from potable reuse plants that process treated wastewater via RO often contain high concentration of nonbiodegradable potentially damaging substances such as pharmaceuticals, per- and polyfluorinated substances, and endocrine disruptors, which are typically not found in seawater discharges.

The potential environmental impacts of ocean desalination should also be weighed against its advantages, particularly with respect to representing a hydrologically independent (drought-proof) new water supply source that can be developed and controlled by local stakeholders, offering a reliable source of high-quality potable water. Where traditional inland surface and groundwater sources may be constrained due to unique regulatory restrictions, typically more complex water rights issues, poor water quality, or supply reductions due to climate variability, ocean desalination offers water supply reliability for local and regional agencies.

Finally, as this chapter reviews key aspects of ocean desalination environmental impacts from a national perspective, it is essential to recognize the site-specific variability of ocean desalination environmental sensitivity, considering the physical site conditions and project-specific design features. More importantly for project implementation, the local regulatory and political framework for ocean desalination varies widely between coastal states where ocean desalination facilities operate or are proposed (Duff et al. 2017). At the national level, the first Trump administration prepared the "Coordinated Strategic Plan to Advance Desalination for Enhanced Water Security," outlining three key goals (NSTC 2019):

- to reduce risk and streamline local planning to support desalination,
- to reduce technical and economic barriers to enable desalination technology usage, and
- to encourage national and international cooperation to innovate and develop desalination technologies.

Other federal actions include making additional ocean desalination funding available through sources such as the Water Infrastructure Finance and Innovation Act and the Desalination and Water Purification Research Program.

## 4.2 GENERAL ENVIRONMENTAL CONSIDERATIONS IN DESALINATION PLANT DESIGN

Most desalination facilities in the United States treat brackish water, and well over half of them are located in the states of Florida, Texas, or California, with varying levels and types of regulatory and environmental oversight in each of these locations. As of 2018, Florida leads the nation with 167 municipal desalination facilities compared to 58 in California and 52 in Texas (US Department of the Interior 2018). Florida's Division of Water Resources Management also notes that "desalination has already been determined to be a feasible and cost-effective supply alternative by water supply utilities" (Florida Department of Environmental Protection 2010). However, while Florida leads the nation in the number and capacity of installed desalination facilities, only three of their existing facilities are treating ocean water, the largest of which is the 25-million gal per day (mgd) (94 ML/d) Tampa Bay Seawater Desalination Plant discussed in Chapter 8.

In Texas, there are currently no municipal desalination facilities treating ocean water; however, a 2002 desalination initiative tasked the Texas Water Development Board (TWDB) with developing the state's first large-scale seawater desalination facility to produce drinking water. Later that year, TWDB launched the Seawater Development Initiative, and in 2015 passed House Bill 4097 to streamline and expedite the regulatory processes associated with seawater desalination (TWDB 2019). At the time of this publication, TWDB was pursuing its first municipal ocean desalination facility in Corpus Christi, with a planned ultimate capacity of 51 mgd (191 ML/d). California currently has more operational desalination facilities treating ocean water than any other state, including the country's largest facility in Carlsbad (see Chapter 8). California's ocean desalination regulatory framework, however, is complex and currently guided by a variety of federal, state, and local regulations depending on project and site specifics. In addition to applicable federal regulations, ocean desalination facilities in California are governed by several state agencies and their associated regulations. A partial listing of California's desalination regulations follows.

- The State Water Resource Control Board and its Water Quality Control Plan for Ocean Waters of California and Desalination Provisions (adopted in May 2015, with the desalination provisions later approved in 2016 and continuing evaluation for further amendments), which guides seawater desalination plant development, permitting, and regulation (California State Water Resources Control Board 2019)
- The California Coastal Commission (2025a) that implements the California Coastal Act (as amended in 2019) and Sea Level Rise Guidance (2018), both of which regulate development along California's coast (California Coastal Commission 2025b)
- The State Lands Commission that manages and leases all state-owned or managed lands for facilities to be sited on
- The Regional Water Quality Control Boards that implement the California Ocean Plan and California Water Code on a project-specific basis, in addition to administering the National Pollutant Discharge Elimination System (NPDES) permit process and related permits and approvals in accordance with the federal Clean Water Act and state laws

It is important to note that all the aforementioned state agencies interact during the California Water Code determination in support of the Regional Water Quality Control Board. When mitigation is needed to offset the impacts of seawater desalination in California, National Oceanic and Atmospheric Administration National Marine+ Fisheries Service (NOAA NMFS), US Fish and Wildlife Service (FWS), and the California Department of Fish and Wildlife must also be consulted, depending on the impact and the mitigation project selected. State-regulated private utilities are further regulated by the California Public Utilities Commission. Ocean desalination projects in California must comply with the National Environmental Policy Act and associated federal regulations when federal permits or funding is involved. Typically, coastal desalination facilities will require federal permits or approvals from the US Army Corps of Engineers for compliance with the Clean Water Act Section 404 (fill or dredge in coastal waters or upland "jurisdictional waters") and the Rivers and Harbors Act regarding structures in navigable waters, which in some cases is interpreted to include subsurface intake (SSI) wells located beneath the ocean floor. Federal permits and approvals also require coordination and consultation with other federal agencies, including the NOAA NMFS, FWS, and State Historic Preservation Officer, in addition to situations where desalination facilities are located on or traverse federal property.

In California, all projects must also comply with the California Environmental Quality Act (CEQA) and a myriad of other state and local regulations. The CEQA process alone can often require several years or more for compliance, with ocean desalination projects requiring as many as 10–20 years to wind through the CEQA and permit approval process. The complexity of California ocean desalination regulations is further complicated by opposition from a variety of stakeholders (on environmental, cost, and other grounds),

**Table 4-2 Cost of water of recent desalination projects in the Middle East and Israel**

| Project | Location | Capacity ($m^3$/day) | First-year Cost of Water (US\$/$m^3$) |
|---|---|---|---|
| Hassyan | United Arab Emirates | 545,000 | 0.306 |
| Sorek 2 | Israel | 548,000 | 0.413 |
| Jubail 3a | Saudi Arabia | 600,000 | 0.440 |
| Yanbu 4 | Saudi Arabia | 450,000 | 0.470 |
| Tawellah | United Arab Emirates | 909,200 | 0.495 |
| Shuqaiq 3 | Saudi Arabia | 450,000 | 0.521 |
| Rabigh 3 | Saudi Arabia | 600,000 | 0.531 |

NOTE: US\$1.0/$m^3$ = US\$3.785/1,000 gal
NOTE: Data have been converted to US metrics.
*Adapted from* WateReuse Association Desalination Committee 2011

in addition to political and litigation risk as projects move through the design and development review and approval processes. For example, the nation's largest ocean desalination facility in Carlsbad was initiated in 1998 and not approved until 2012 and then survived numerous separate lawsuits, finally beginning water production in December 2015. The 50-mgd Huntington Beach Desalination Project was initiated in 2001, approved by the State Lands Commission in 2017 after four separate environmental impact reports, and is currently still in the litigation and permit approval process. California American Water's Monterey Peninsula Water Supply Project application was submitted to the Public Utilities Commission (PUC) in 2004, eventually approved by the PUC in 2018, and is still pending permit approvals.

Differences in environmental regulations worldwide have made a significant impact on desalination costs. Table 4-2 shows first year costs of desalination projects in the Middle East and Israel, which are several times lower than seawater desalination plants in California and Australia, where new seawater desalination projects have undergone increased environmental scrutiny. This cost differential exists despite the fact that the design, intake and outfall types, and configurations of the plants in California and Australia are similar to those in the plants listed in Table 4-2. Much of the cost differential can be attributed to project development costs, with environmental studies and permitting adding significantly to project development. Intake approaches can also add considerably to project costs, with the long distance tunneling used for Australian intakes and outfalls far more costly than the simpler intakes and outfalls used at Middle Eastern facilities. Further discussion on the impacts of design approaches on facility cost is provided in Chapter 5.

Most countries worldwide do not have regulatory requirements specifically pertaining to the operation of desalination plants and minimization of their environmental impacts (Mickley and Voutchkov 2016). However, internationally, most facilities are built under nonrecourse project financing, which results in lenders employing the Equator Principles. In the simplest terms, the Equator Principles require that projects be audited to ensure that environmental best practices and other best practices are consistently incorporated. In this way, the absence of specific local regulation does not necessarily mean that best practices are not being utilized.

In May 2015, California introduced the first regulations in the world specifically pertaining to environmental impacts of seawater desalination plant intakes and outfalls (SWRCB 2015). These regulations incorporate very stringent environmental requirements and mitigation measures, which aim to promote the use of SSI wells instead of open seawater intakes and to significantly increase the environmental mitigation requirements for desalination plant operations. Such stringent regulations are projected to increase the cost of production of desalinated water by 20–25 percent, and since their introduction, such regulations have delayed the implementation of some new full-scale desalination projects in California.

Desalination plant-specific intake- and outfall-related regulations have not been adopted by any other US state or other country worldwide so far, including the countries of the Middle East region. The approach taken by most countries on the environmental regulation of desalination plant intakes and outfalls has been to incorporate in new projects the best-practices and "lessons learned" from over 50 years of experience in the desalination field and for regulators to request incorporation of such best practices through the environmental project review process, rather than to mandate such practices by legislation or to require specific mitigation measures that result in direct increase of the production costs of desalinated water. Lack of clear regulatory guidance at a state level and regulations tailored to penalize desalination have resulted in the slowdown of the implementation of large desalination projects in California and some other regions of the world.

Many countries outside of the Middle East and Israel—especially Australia, the United States, and European nations—do have nongovernment organizations (NGOs) representing environmentalists' concerns, which exert significant influence on the environmental review process and legislation pertinent to desalination projects. Reduction in the size of the projects, or in some cases introduction of location and environmental mitigation conditions for the projects, has resulted in measurable increases in the desalination project construction and operation costs and ultimately of the cost of water.

The 50-mgd Claude "Bud" Lewis Carlsbad Desalination Plant in California (US$490 million in capital costs, US$55.5 million in annual operations and maintenance [O&M] costs, and cost of water of US$2,060/acre-ft) is another example of how responding to very stringent social and environmental safeguard regulations—including extensive studies to evaluate potential environmental impacts—has resulted in a significant increase in the desalination project cost. The project went through an 8-year permitting process that involved numerous environmental studies and lawsuits from environmental NGOs, all of which were dismissed in legal proceedings and have resulted in project development expenditures of over US$60 million—approximately half the capital costs of the slightly higher-capacity desalination project in Al Dur, Bahrain, which coincidentally uses the same type of intake, pretreatment, and desalination technologies and configuration as Carlsbad. The capital cost of the Al Dur project is approximately five times lower than that of the Carlsbad SWRO plant.

Besides the US$60 million of environmental review related expenditures, the Carlsbad project costs include US$28 million in capital expenditures for mitigation of environmental impacts from the intake operation and over US$50 million in expenditures associated with incorporating carbon footprint mitigation measures. In addition, these capital costs include the expenditures for successfully defending the project against the aforementioned lawsuits, which added a total of US$4 million of additional capital costs to the project. It is not clear whether the increased environmental scrutiny employed at Australian and US desalination facilities has had a discernible impact on the environmental footprint of the desalination plants; however, several long-term studies looking at impacts have been published, noting that impacts have been minimal (Avramidi et al. 2022; Kelaher et al. 2022; Clark et al. 2018).

## 4.3 SOURCE WATER INTAKES

The main purpose of intakes is to collect source water of adequate quantity and quality needed to produce desalinated water. Because intake water quality has a significant impact on desalination facility operations, desalination intake design should target collection of water with minimal inorganic (to the extent possible) and organic content, including marine life. Seawater desalination facilities use two general types of seawater intake facilities: subterranean or SSIs (wells and infiltration galleries) and open (or "screened") ocean intakes. It should be noted that a subsurface infiltration gallery (SIG) will typically operate like a subsurface well intake with respect to eliminating intake-related impingement and entrainment (I&E) issues. However, it may face similar or greater impact during construction as an open intake because construction is typically done offshore and because of the large area needed for this intake. Furthermore, SSIs are limited in their total intake volume capacity unlike surface intakes. Note that currently there are no operating SIG intakes in the United States, although a beach SIG was tested in Long Beach (Missimer et al. 2013). Worldwide, SSIs are used mainly for plants of capacity of 1 mgd or less (less than 4 ML/d). Less than 1 percent of the desalination plants with capacity of 5 mgd (19 ML/d) or more have SSIs. At present, all desalination plants worldwide with capacity of 25 mgd (95 ML/d) or more have open intakes only.

I&E of marine organisms are a key consideration for ocean desalination facilities, an issue that is primarily associated with open ocean intakes. Impingement occurs when aquatic organisms are trapped against intake screens by the velocity and the force of the flowing source water. Entrainment occurs when marine organisms pass through the intake screens and enter into the process equipment and treatment facilities where some marine organisms are damaged or destroyed.

The impacts of I&E vary considerably with the intake volume and velocity of the feed seawater and the use of design features or mitigation measures developed to minimize their impact. I&E of aquatic organisms are not environmental impacts unique to open intakes of seawater desalination facilities. Conventional freshwater open intakes from surface water sources (e.g., rivers, lakes, estuaries) may also cause I&E, as in the State Water Project intake pumps in California's Bay-Delta.

Similar to environmental impacts from other aspects of desalination facility operation, the magnitude of impacts due to I&E varies significantly from one location to another. Therefore, when assessing the impacts caused by the intake of a desalination facility, it is essential to consider the applied technology and operational practices, the actual volume and velocity of water being drawn into the desalination facilities, and the species composition and abundance of the seawater surrounding the intake. As discussed in the next section, SSIs essentially eliminate I&E impacts, at least those associated with source water intake. In California, SSI has been determined by the state to be the preferred desalination source water intake technology, with only screened ocean intakes allowed if the ocean desalination proponent can demonstrate that SSI is not feasible (with screen spacing not greater than 1 mm and intake velocity not greater than 0.5 ft/s (0.15 m/s), subject to additional data collection, mitigation, and monitoring requirements pursuant to California's Ocean Plan Desalination Provisions Coastal Act, and other regulations).

## 4.3.1 Subterranean or Subsurface Intakes

SSIs, although generally viewed as environmentally preferable to screened (or open) ocean intakes, have the potential to cause their own range of environmental impacts like loss of coastal habitat during construction, disruption of nearshore groundwater and coastal wetlands, and temporary or permanent loss or disruption of coastal recreational lands, as well as visual and aesthetic problems. The magnitude of these impacts and potential avoidance, minimization, and mitigation measures are discussed in the following sections.

**I&E.** Because SSIs naturally filter the collected seawater at low velocities through the granular formations of the coastal aquifer in which they operate, their use eliminates entrainment of marine organisms into the seawater desalination facility. The source seawater collected by this type of intake typically does not require mechanical screening, and therefore, SSIs do not impinge marine organisms in the area of the intake. Some agencies require evaluation of discharge entrainment and potentially hypoxic conditions associated with anaerobic seafloor conditions.

**Visual and aesthetic impacts and mitigation measures.** The visual and aesthetic impacts of SSI are dependent on the location of the wellhead and the specific wellhead type, design, and construction methods. If the SSI can be constructed below grade, at grade, or near grade to minimize impacts, submersible well pumps can be installed below grade and the structures made watertight. The electrical controls and auxiliary equipment can be installed within the watertight structure or sited at a remote location near the intake off the beach for protection. In these cases, there may be little or no visual or aesthetic impact for this kind of SSI design, other than temporary impact during construction. In some cases, intake wells are installed above ground because their installation in underground structures increases their costs and complicates their maintenance.

If the beach intake is constructed above grade, as typically is the case, the magnitude of this impact will vary according to the physical placement of the wells and the height above grade that is required. With radial collector wells, such as a Ranney well system, it is possible to locate the well structure back from the beach and extend the well screens out underneath the beach to reduce visual impacts.

Considering that the desalination facility source water must be protected from acts of vandalism and terrorism, the individual above-grade SSI facilities may have to be fenced off or otherwise protected from unauthorized access.

The SSI structure is essentially a concrete vault that must have secured access and/or be fenced off, which could damage coastal visual and aesthetic appeal. Subgrade or near-grade completion could utilize secured access hatches and would have limited long-term visual impacts. Because coastal areas are commonly visually sensitive areas, the installation of above-grade beach wells may affect the recreational and tourism value of the seashore and may change the coastal appearance and character.

Installing the SSI wells and pumping gallery in a location set back from the active coastal recreation areas is usually preferable, especially if a less environmentally sensitive area of adequate size is available near the coastline. These two general locations for the wells can utilize different well designs to accommodate local geographic settings and other socio-environmental issues. These designs include wells that are

- completed below grade, which can include the wellhead installed in a vault buried completely to eliminate visual impacts, either on or behind the beach;
- completed at or near grade with only minimal surface features to provide low visual impacts for locations in residential areas and for public use;
- completed above grade, especially where the top of the well structure needs to be above known or anticipated flood elevations to allow access during high water events on or behind the beach; and
- completed with a dual design, whereby the well portion of the system can be located closer to the source water (e.g., out on the beach), and the pumping portion of the system can be located further back from the source water to minimize noise and visual impacts and connected with the SSI wells via underground piping.

It should be noted that the use of the more environmentally palatable intake well configurations may result in a significant (50–200 percent) increase in the overall cost of intake construction. In any case, the wellheads have to be installed so that they are accessible for inspection and maintenance, taking into consideration that well pumps for plants over 5 mgd (19 ML/d) are usually large and heavy and cannot be lifted manually if inspections, maintenance, or repairs are needed.

**Loss of coastal habitat during construction: impacts and mitigation measures.** Smaller seawater desalination facilities (i.e., less than 5 mgd) typically require a limited number of SSI wells, and their impact on the coastal habitat during construction is generally correspondingly reduced compared to larger desalination facilities (i.e., 5 mgd or greater). These lower-capacity SSI systems can often be constructed as low-profile structures to minimize visual impacts. However, even with very small coastal desalination facilities, site-specific conditions including regulatory permitting requirements can make project implementation costly, delayed, or canceled altogether. In California, for example, the California American Water's community-sized Monterey Peninsula Water Supply Project, an SSI desalination facility at 6.9 mgd (26.1 ML/d), has seen multiple delays and litigation for a variety of reasons over its 15+ year history of permitting, and has most recently been recommended for denial for the second time by the California Coastal Commission.

Because of the higher number of wells needed to supply adequate amounts of water for a large seawater desalination facility, construction of these facilities may result in impacts over a larger area of coastal habitat, and where the associated SSI facilities involve above-grade structures, they would generally have increased environmental impact on coastal habitat and the visual and aesthetic environment. During SSI siting and design, it is particularly important to avoid impact on species or habitats that have federal or state protection. In California, sensitive coastal habitat is referred to as environmentally sensitive habitat areas (ESHA), which is a broadly defined and applied term that could include wetlands, natural habitat, trees providing nesting or foraging opportunities, coastal sand dunes, or other sensitive areas.

**Coastal wetland habitat: impacts and mitigation measures.** Any intake wells that are operated in coastal areas will likely have at least some impact on local groundwater resources and other features like perched water, wetlands, or saltwater–freshwater interfaces like coastal lagoons, as the hydraulic influence from SSI pumping in the area may affect water levels and alter groundwater flow gradients. Special attention should be given to seawater intake well siting in the vicinity of existing coastal wetlands and other groundwater users to evaluate these hydraulic impacts. The operation of large intake wells located adjacent to coastal wetlands may result in a drawdown of the groundwater table that could increase salinity, depending on site-specific well design and hydrogeologic conditions. Potential mitigation measures in this case are installation of a higher number of smaller capacity wells where the radius of influence does not reach the wetlands or the use of a radial collector or Ranney well–type configuration. Other possible solutions are the use of slant wells, which have a much lower drawdown of coastal groundwater, or Neodren shallow horizontal directionally drilled wells. While there are Neodren SSI wells operating only in Spain where the technology was developed, slant wells have been successfully tested at two locations in California and are currently in the permitting process for full-scale facilities (proposed at the Doheny Ocean Desalination Project in Dana Point and the Monterey Peninsula Water Supply Project in Marina). Out of the over 250 desalination plants in Spain, very few use slant wells, and none of them has capacity larger than 15 mgd (57 ML/d).

Similar concerns and solutions can be applied to conditions where the radius of influence of the intake wells extends to the area of a landfill or other contaminated site (e.g., leaking fuel storage tanks of a gas station) located near the coast. In this case, the SSI could immobilize hazardous compounds contained in the waste site and contaminate the water source. Numerous reports (e.g., ISTAP 2015) prepared for screened ocean intake facilities have identified various other feasible concerns with implementing SSI, especially with larger ocean desalination facilities.

**SSI operation: permanent impacts and mitigation measures.** The permanent construction-related impacts are mainly associated with the excavation and disposal of sand and other materials from the shoreline in order to drill and develop the intake wells.

For most SSI, following construction, the long-term operational impacts of the SSI system are generally fairly limited, with occasional maintenance access required via the surface access vault. Intake wells, while generally requiring minimal maintenance, may periodically require reconditioning or abandonment in some cases. In the event of abandonment, a new well would need to be drilled. An infiltration gallery would have the highest impact on the ocean bottom habitat because its construction likely involves substantial temporary structures and excavating a large area of ocean bottom (at a depth of 10 ft or 3 m or more) and replacing it with artificial sand and gravel. The area of the ocean bottom habitat that will be removed and destroyed may be significant, especially for construction of intakes for large seawater desalination facilities. In addition, ongoing periodic need to rehabilitate or replace substrate would create long-term marine impacts associated with removal and replacement or repair activities.

One important impact that has occurred when including infiltration galleries and other wells is loss of filtration substrate due to beach erosion or heavy storms. Because such events are very difficult to control, regulatory agencies issuing drinking water permits often do not recognize SSIs as filtration systems for pretreatment of the collected seawater and do not award pathogen log-removal credits for SSIs that they will usually award to conventional man-made pretreatment filters.

Additional consideration based on experience in California and elsewhere is that as compared to open intakes, which have a useful life of 50 years or more, the useful life of most coastal wells is only 10–15 years, which means that the environmental impacts associated with the construction of such intakes will be doubled over the useful life of a typical desalination plant (25–30 years). Potential environmental impacts of SSIs and operational implications are discussed in greater detail elsewhere (Voutchkov 2013, 2012).

**Additional environmental considerations for SSI siting and design.** Various other environmental factors should be considered in the siting, design, and environmental permitting process for a proposed SSI facility. These include but are not limited to the following issues unique to coastal desalination facilities.

*Coastal recreational lands.* SSI may impact public or private coastal recreational lands, requiring additional permitting as well as special consideration for seasonal construction and measures to minimize effects upon coastal access, viewsheds, parking, and recreational spaces.

*Cultural resources.* SSI may have a higher potential for archaeological resource sensitivity due to proximity to the coast.

*Coastal erosion and sea-level rise.* SSI may be more susceptible to coastal erosion and sea-level rise, requiring siting outside of composite 100-year flood hazard zones in light of projected storm surge, tsunami, high tide, and sea-level rise conditions. In California, this includes specific protocols adopted by the California Coastal Commission as recently amended by the Ocean Protection Council in its February 2020 Strategic Plan, which recommends that agencies consider a minimum of 3.5 ft (1.1 m) of sea-level rise by 2050 during design and permitting of coastal facilities (California Ocean Protection Council 2020). SSI well siting should consider coastal erosion and sea-level rise vulnerability, limit aboveground structures within the affected areas, and provide for maintenance access to any buried structures. SSI construction may be particularly challenging in locations subject to coastal wave action.

### 4.3.2 Screened Ocean Intakes

Similar to SSIs, open ocean intakes would have environmental impacts associated with their construction and operation.

**I&E impact and mitigation measures.** As indicated previously, I&E are considered the two most significant environmental issues of open ocean intakes or screened ocean intakes; the latter term simply refers to the current practice of requiring advanced marine life design measures for open ocean intakes, which in California must be no larger than 1-mm screen spacing. Impingement rates from a desalination facility open ocean intake depend on the intake design, location, and velocity of the feedwater. Impingement mortality of marine species is typically caused by exhaustion or injury due to the suction force of the water, being impinged alive on traveling water screens and removed from the water, or from the physical force of the water jets used to clear the screens if a fish return system is used to remove live fish from the intake and return them to the source water. Impingement mortality is largely eliminated by the use of small-mesh screens on the intake structure designed with a maximum 0.5 ft/s (0.15 m/s) through-screen velocity. At this intake velocity, most fish can sense, evade, or escape the feedwater intake.

While specific intake design may reduce or eliminate impingement, all desalination open water intake systems will cause a certain degree of entrainment. Entrainment impact is associated with marine species mortality caused by the equipment, chemicals, or treatment facilities used for water treatment as the marine organism is carried through the source water intake system.

Entrainment impact is typically proportional to the volume of source water collected by the intake and varies widely based on the amount of seawater required by the facility, intake velocity, location, depth, existing biological conditions of the affected area of the intake structure, and the intake technology/equipment used. To predict and assess impacts from a desalination facility intake, site-specific studies are necessary to identify habitats and species in the area that might be vulnerable to I&E.

The methods for mitigation of I&E of marine organisms can be divided into three categories: alternative open intake technologies, operational impingement reduction measures, and impact mitigation measures.

**Alternative desalination facility open intake technologies.** Table 4-3 presents a number of technologies that are classified based on biological effectiveness (i.e., ability to achieve significant reductions in I&E). The feasibility of these technologies for the site-specific condition of a given desalination project should be evaluated based on the following criteria:

- ability to achieve a significant reduction in I&E for all species, considering variations in abundance of all life stages;
- feasibility of implementation at the desalination facility;
- cost of implementation (including installed costs and annual O&M costs); and
- impact during desalination facility operations.

**Operational measures.** Operational mitigation measures are used to reduce the amount of flow and velocity of entrance of the source water into the desalination facility intake to minimize I&E of marine organisms. Operational measures may consider reduction of plant intake flow during certain periods of the day (typically at night) or periods of the year (typically during the spring and summer months) when the concentration of marine species in the source water is at its highest level. Plant intake flow may be reduced by either reduction of desalination facility overall freshwater production yield or by operating the desalination facility at higher recovery (Voutchkov 2013).

Entrainment of marine organisms is mainly proportional to intake flow. Installation of variable frequency drives (VFDs) on the intake pump motors could reduce the flow that enters the desalination facility by collecting only as much flow as needed at any given time, meeting the desalination facility freshwater production targets. The Carlsbad desalination facility was successfully re-permitted in May 2019 at an increased production capacity (to 56,000 acre ft/year, 50 mgd, or 190 ML/d) with the requirement of installing fish-friendly pumps (San Diego County Water Authority 2020) and to design and test 1 mm slot–width cylindrical wedge-wire intake screens that could eliminate impingement and reduce entrainment.

**Table 4-3 Potential impingement/entrainment (I&E) reduction technologies**

| Technology | Impact Reduction Potential | |
|---|---|---|
| | Impingement | Entrainment |
| Modified traveling screens with fish return | Yes | No |
| Replacement of existing traveling screens with fine mesh screens with fish return | Yes | Yes |
| New fine mesh screening structure with fish return | Yes | Yes |
| Cylindrical wedge-wire screens, fine-slot width | Yes | Yes |
| Fish barrier net | Yes | No |
| Fine mesh dual flow screens with fish return | Yes | Yes |
| Angled screen system, fine mesh with fish return | Yes | Yes |
| Behavior barriers (e.g., light, sound, bubble curtain) | Maybe | No |
| Variable speed drives | Yes | Yes |

*Source:* USEPA 2014

**Impact mitigation measures.** In addition to the implementation of technological and operational measures to minimize I&E impacts, the impact of screened ocean intakes on the surrounding aquatic environment can be further mitigated by implementing projects aimed to preserve, restore, or enhance this environment by creating additional habitat for species in kind to the impacted marine organisms.

Mitigation projects to be considered should target the preservation, restoration, creation, or enhancement of a coastal habitat comparable to that impacted by the intake. In many cases, however, the coastal habitat impacted cannot be preserved, restored, created, or enhanced. In these cases, out-of-kind mitigation should be evaluated. Out-of-kind mitigation involves the preservation, restoration, creation, and/or enhancement of habitat in the area that will ultimately improve the overall coastal ecosystem to a level comparable with that before the impact occurred. Key eligibility criteria for such mitigation projects may include consistency with the applicable requirements of federal, state, and local agencies that have jurisdiction over coastal habitat restoration actions as well as restoration of marine habitat similar to the marine habitat affected. Since impacts of open intakes are primarily to "open water" habitat, ocean desalination projects may require development or compliance with methodologies for and scaling of out-of-kind mitigation and effectiveness testing. These methodologies should be considered for

- projects located in close vicinity of and preferably in the watershed near the intake,
- projects that hold the promise for long-term environmental enhancement benefits, and
- projects that have opportunities for leveraging of funds/availability of matching funds.

Examples of types of mitigation projects include

- wetland restoration,
- coastal lagoon rehabilitation,
- reconstruction of historic sediment elevations to promote reestablishment of eelgrass beds,
- marine fish hatchery enhancement,
- contributions to a marine fish hatchery stocking program,
- artificial reef development, and
- kelp bed improvement.

Selection of the most suitable mitigation measures would need to be completed based on a life-cycle cost–benefit analysis.

**Screened ocean intakes: construction impacts and mitigation measures.** Screened ocean intakes can generally be divided into two types—onshore and offshore. Construction of onshore screened intakes (such as at the edge of a bay, inlet, or in the case of the Carlsbad desalination facility, a coastal lagoon) involves minimum disturbance of marine life in the vicinity of the intake, but they are often highly visible structures with potential impact on coastal aesthetics. Onshore intakes may also have coastal habitat or aquatic species concerns, depending on the location.

The preferred option for offshore intakes is to utilize an existing offshore pipeline, whether that be a coastal power plant intake structure or a coastal intake or discharge structure for a wastewater or other industrial facility (as was the case for the City of Santa Barbara's Charles E. Meyer Seawater Desalination Plant [see Chapter 8]). If an existing ocean structure is not available, construction of a new offshore intake would require extensive environmental studies and a much higher degree of permitting complexity and uncertainties and may involve considerable regulatory agency and stakeholder opposition. Offshore intakes are typically constructed by installing the intake pipeline in one of several methods: installing directly on the surface of the ocean bottom and securing the pipeline with weighted blocks, installing the intake pipeline in an excavation trench, directional drilling of the intake pipeline/tunnel under the ocean floor, or some combination of the above. Intake pipeline installation in a trench excavated from the ocean bottom usually is the most environmentally intrusive. Therefore, if the intake area contains environmentally sensitive habitats, the preferred method of intake pipeline installation is directional drilling at least 5–15 ft (1.5–4.5 m) under the ocean floor, subject to coastal erosion analysis to evaluate predicted future scour depths in the nearshore environment.

While the onshore open intake is lowest in cost, it is the most visible structure and in closer proximity to nearshore sensitive environments, and often for this reason, it is avoided. An intake structure drilled under the ocean floor is the most costly and complex type of such facility but has the advantage of minimal disturbance of the ocean flora and fauna during construction. In California, several large ocean desalination facilities have considered using new drilled ocean intakes, including the Deepwater Desalination Project in Moss Landing.

## 4.4 CONCENTRATE DISCHARGE CONSIDERATIONS

One of the key limiting factors for the construction of new desalination facilities is the availability of suitable conditions and locations for disposal of the desalination discharge concentrate or concentrate stream.

### 4.4.1 Introduction

Concentrate is generated as a byproduct of the separation of the minerals from the source water used for desalination. This liquid stream contains most of the minerals and contaminants of the source water and pretreatment additives in concentrated form. The concentration of minerals and contaminants in the concentrate from seawater desalination facilities is usually 1.5–2.5 times that in the source water, depending on the recovery of the desalination facility. If chemical pretreatment is used—such as coagulants, antiscalants, polymers, or disinfectants—some or all of these chemicals may be disposed of along with the plant discharge concentrate.

The quantity of the concentrate is largely a function of the plant recovery, which in turn is highly dependent on the salinity concentration of the source water. Seawater desalination facility recovery is typically limited to 40–65 percent. The salinity level of concentrate from seawater desalination facilities usually is in a range of 65–85 parts per thousand (ppt) (NOTE: 1 ppt is equal to 1,000 mg/L) to 85,000 mg/L, while that from brackish plants may vary between 1.5 and 25 ppt, depending on source water salinity. The number of particles, total suspended solids (TSS), and biological oxygen demand (BOD) in the concentrate is usually below 5 mg/L because these constituents are removed by the plant's pretreatment system. However, if plant pretreatment waste streams are discharged along with the concentrate, the blend may contain elevated turbidity, TSS, and occasionally BOD. Acids and scale inhibitors added to the desalination facility source water will be rejected in the concentrate and will impact its overall mineral content and quality. Often scale inhibitors contain phosphates or organic polymers.

Because membranes are more permeable to some chemicals than others, variable concentration factors may apply for specific chemicals. Exactly how the brine concentration factor impacts the disposal of concentrates depends heavily on the means of disposal. In some cases, volume minimization (high concentration factor) will be preferred, whereas in cases where the concentrate is to be discharged to waterways, low concentration may be more important than low volume.

For example, the Perth Seawater Desalination Plant in Australia is a two-pass RO plant operating with a first-pass recovery of 45 percent and a second-pass recovery of 90 percent. This corresponds to an overall concentrate concentration factor of approximately 1.7 times. Based on a source water TDS of 33,000–37,000 mg/L, the plant produces an overall RO concentrate TDS of approximately 65,000 mg/L.

With most seawater desalination facilities producing concentrate 1.5–2 times more concentrated than ambient seawater, the concentrate may have a negative impact on the aquatic environment in the area of the discharge. This impact is very site specific and depends mostly on the ambient mixing forces (currents, wave energy, etc.) and the salinity tolerance of the specific marine organisms inhabiting the water column and benthic environment influenced by the discharge. The existing US Environmental Protection Agency (USEPA) whole effluent toxicity (WET) tests are indicative of the level of salinity that causes mortality of preselected test organisms, which may or may not inhabit the discharge area. WET testing is an important element of the comprehensive evaluation of the effect of the concentrate discharge on the aquatic life. Completion of both acute and chronic toxicity testing is recommended for the salinity levels that may occur under worst-case combination of conditions in the discharge (Mickley and Voutchkov 2016).

### 4.4.2 Concentrate Discharge Impact on the Environment

Concentrate from seawater desalination facilities using open ocean intakes generally has the same color, odor, oxygen content, and transparency as the source seawater from which it was produced, and an increase or decrease in salinity will not change its physical characteristics or aesthetic impact on the environment.

There is no relationship between the level of salinity and biological or chemical oxygen demand of the desalination facility concentrate. More than 80 percent of the minerals that encompass concentrate salinity are sodium and chloride, and they are not a prime food source or macro- or micronutrients for aquatic organisms.

Salinity contained in concentrate discharges from seawater desalination facilities is not of anthropogenic origin as are the pollutants contained in discharges from industrial or municipal wastewater treatment plants or water reclamation plants. The minerals contained in the seawater desalination facility concentrate discharge originated from the same source to which they usually are returned. As a result, the environmental effect of seawater desalination on the ocean is somewhat equivalent to the effect of naturally occurring evaporation. In addition, although the seawater desalination concentrate has higher salinity than ambient ocean water salinity, these elevated salinity levels are limited to the immediate area at the point of discharge due to rapid dilution from discharge jets or natural ocean mixing. Furthermore, salinity levels quickly return to ambient ocean water salinity levels within a relatively short distance of the discharge (this varies depending on site-specific details and requires hydrodynamic modeling to ascertain).

Naturally occurring evaporation tends to concentrate salinity in shallow nearshore ocean embayments during the high-temperature dry periods of the year, and they are diluted during the rainy periods of the year, keeping a net-zero salinity effect. Similarly, seawater desalination facilities temporarily remove a small portion of ocean water in producing fresh drinking water, which in turn may be returned to the ocean via the ocean discharges of the wastewater treatment plants located in the vicinity of the desalination facility. Even at locations where extensive water reuse projects are utilized, a portion of the water will almost universally be returned to the ocean with salinities lower than the background salinity in the ocean.

The state-of-the-art desalination processes employed in contemporary desalination plants use a very limited number of chemicals. All chemicals added in various treatment processes of desalination plants are of food grade quality, biodegradable, and specifically selected not to cause aquatic life toxicity. Therefore, the discharges from seawater desalination plants are neither toxic nor harmful for marine life and are engineered to dissipate rapidly and without permanent alterations to the surrounding marine ecosystem (IDA Energy and Environment Committee Forum 2019).

Over the past 5 years, many countries with large desalination plants like Saudi Arabia, Australia, Israel, and Spain have initiated comprehensive programs for green desalination that aim to reduce both the number and the types of chemicals used in the production of desalinated water. These programs will ultimately convert all existing facilities into chemical-free seawater desalination plants by implementing the latest advancements of desalination technology and science.

Desalination plants used to continuously chlorinate their intake seawater using sodium hypochlorite to suppress the growth of marine life in the intake piping and on the RO membranes. Such practice was abandoned by most desalination plant operators close to a decade ago, and currently chlorination is used only once or twice per month for a period of 6–8 h unless there is high turbidity organic content (e.g., algae, fungi, bacteria) in the intake seawater. In addition, some desalination plant operators do not apply any disinfectants to the intake seawater because they prefer to use the pretreatment system of the plant for control of biofouling instead of chemicals. Before discharge, the concentrate is typically dechlorinated to meet applicable ocean water quality standards.

Ferric chloride and ferric sulfate are the most commonly used coagulants for pretreatment of seawater at present. These chemicals used to be added at a constant rate and a relatively high dosage. At present, the desalination industry has adopted real-time monitoring of the content of solids in the seawater and automated adjustment of the coagulant dosage proportionally to the actual content of suspended solids in the water. This operational strategy, introduced over the last 10 years at most plants worldwide, has reduced the use of coagulants to less than 50 percent of what it once was.

Acids and flocculants were used for optimization of the chemistry of water treatment in many desalination plants until a decade ago. Most modern advanced desalination plants and skilled plant operators today no longer use acids and flocculants for pretreatment; instead, they rely on optimized pretreatment system design and operation to manage water chemistry.

At one time, antiscalants and sodium hydroxide were commonly applied in many desalination plants worldwide, mainly to prevent scaling associated with removal of boron from the desalinated water. Since 2014, when the World Health Organization increased the drinking water guideline limit for boron from 0.5 mg/L to 2.4 mg/L, desalination plants have typically discontinued the addition of sodium hydroxide and antiscalants.

The desalination industry as a whole is constantly developing and adopting new chemical-free, renewable energy–based technologies. One such research area is to use chemicals extracted from the brine for posttreatment of the desalinated water instead of using commercially supplied calcium compounds such as lime from limestone.

### 4.4.3 Use of Desalination Brine for Carbon Sequestration

One area of developing interest is in the use of desalination brine for enhancing carbon sequestration within the ocean. Ocean alkalinity enhancement (OAE) has been proposed as a method for reducing greenhouse gases (GHGs) from the atmosphere in an attempt to reduce the impact on climate change (Hartmann et al. 2023; Renforth and Henderson 2017). A study led by the Massachusetts Institute of Technology–Woods Hole Oceanographic Institution Joint Program estimated that most marine locations reach an uptake-efficiency of 0.6–0.8 mol of carbon dioxide ($CO_2$) sequestration per mole of alkalinity added to the ocean (He and Tyka 2023). While many seawater desalination plants employ ferric chloride or sulfuric acid in their pretreatment processes to lower the pH and alkalinity of the water, a facility that avoided pH suppression while operating at an RO recovery of 50 percent would be expected to have 65–75 percent higher alkalinity in the concentrate than in the ocean water supply. Based on an assumed carbon dioxide uptake efficiency of 0.7 mol of carbon dioxide per mole of alkalinity and an average seawater alkalinity of 128 mg/L, the brine discharge could be expected to reduce atmospheric carbon dioxide by 231 lb for every 1 million gal (27.2 kg/million L) discharged. While carbon sequestration has not historically been a consideration for desalination brine outfalls, minor changes to plant operations could result in significant improvements in the carbon dioxide uptake potential of the brine and the GHG footprint for the overall facility.

### 4.4.4 Salinity Tolerance of Marine Organisms

Environmentally safe disposal of the concentrate produced at seawater desalination facilities is one key factor in determining the viability, size, and costs of a given project. The maximum total dissolved solids (TDS) concentration that can be tolerated by the marine organisms living in the desalination facility outfall area is defined as a salinity tolerance threshold and depends on the type of the aquatic organisms inhabiting the area of the discharge and the period of time these organisms are exposed to the elevated salinity (Phillips et al. 2012) These conditions are very site specific for the area of each desalination outfall, and therefore, it is very difficult to determine the salinity tolerance threshold.

Marine organisms have varying sensitivity to elevated salinity. Some marine organisms are osmotic conformers, meaning that they have no mechanism to control osmosis; therefore, their cells conform to the same salinity as their environment. A large increase in salinity in the surrounding marine environment due to concentrate discharge can cause water to leave the cells of these organisms, which eventually leads to cell dehydration that can result in cell death.

Osmotic regulators are marine organisms that can naturally control the salt content and thereby control the osmotic potential within their cells despite variations in external salinity. Most marine fish, reptiles, birds, and mammals are osmotic regulators and employ a variety of mechanisms to control osmosis. Salinity tolerances of marine organisms vary, but few shellfish (scallops, clams, oysters, mussels, or crabs) or reef-building corals are able to tolerate salinities greater than 40,000–45,000 mg/L (Phillips et al. 2012).

Concentrate disposal may also have impacts other than direct changes in salinity. In some circumstances, concentrate plume density may lead to increased stratification reducing vertical mixing (Van Senden and Miller 2005). This may reduce dissolved oxygen levels in the water column or bottom of the ocean in the area of the discharge, which may have ecological implications. For example, stratification was raised as a particular concern during the planning and assessment for the Perth Seawater Desalination Plant discharge into Cockburn Sound in Australia. However, detailed modeling and site investigation concluded that the anticipated concentrate discharge was unlikely to contribute to the exacerbation of low-oxygen conditions in this case (D.A. Lord & Associates 2005). An ongoing dissolved oxygen monitoring program has been instituted since commissioning of the plant in 2006, and the data to date indicate that the desalination facility discharge has no measurable impact on the dissolved oxygen concentration in the area of the discharge or in Cockburn Sound as a whole. Although a drop in dissolved oxygen was observed, it was found that this change was not linked to the concentrate outfall.

Two studies conducted by the National Marine Science Centre and Marine Ecology Research Center in New South Wales, Australia, found that the abundance of vertebrate fish increased by nearly 280 percent in the area of the Sydney Desalination Plant brine outfall during periods where brine was being discharged (Kelaher et al. 2022; Kelaher et al. 2020). These included "pelagic and demersal fish, as well as fishes targeted by recreational and commercial fishers" (Kelaher et al. 2020). During periods when brine was not being discharged, fish populations returned to levels similar to those before the desalination facility went on line. Similar to fish abundance, fish functional richness was also found to increase, and this increase was maintained even during periods where the brine discharge was offline (Kelaher et al. 2022). While these studies indicated potential positive impacts on vertebrate fish populations, more sensitive invertebrate populations were found to have been negatively impacted within 125 m (137 yds) of the area immediately surrounding the outfall (Kelaher et al. 2022).

A comprehensive study on the effect of the disposal of seawater desalination facility discharges on near-shore communities in the Caribbean was completed in 1998 by the Southwest Florida Water Management District and the University of South Florida (Hammond et al. 1998). This study undertook a detailed analysis of the environmental impacts of the discharges from seven existing seawater desalination facilities in the Caribbean with plant capacities between 0.05 mgd (0.2 million L per day [ML/d]) and 1.6 mgd (6 ML/d) and discharge salinities of 45,000 mg/L–56,000 mg/L. All of the plants use SWRO technology for salt separation and had been in operation for at least 4 years before the study was completed. The study found no statistically significant impact of the desalination facility discharges on the benthic marine life, seagrass, microalgae, and micro- and macroinvertebrates inhabiting the area of the discharge.

### 4.4.5 Additional Considerations for Concentrate Discharge

Ocean desalination design, siting, and permitting considerations may also include other factors, depending on site-specific conditions and applicable regulatory agency requirements. These include but are not limited to the following:

- hypoxic conditions: low oxygen content of concentrate discharge and potential effects on marine organisms;
- turbulence-induced shearing mortality: the potential for concentrate discharge to harm marine organisms entrained in the discharge (In California, this analysis requires a formal protocol as part of the Ocean Plan Amendment consistency evaluation, which is part of any ocean desalination project's NPDES permitting with the regional water quality control boards.);
- effects on nearby marine protected areas: regulatory agencies may require evaluation of concentrate discharge effects on any nearby marine protected areas; and
- future availability of wastewater for blending: where the concentrate discharge is blended with an existing wastewater treatment outfall, regulatory agencies may require discharge modeling to consider future conditions where wastewater outfall flow may be reduced as water reuse increases.

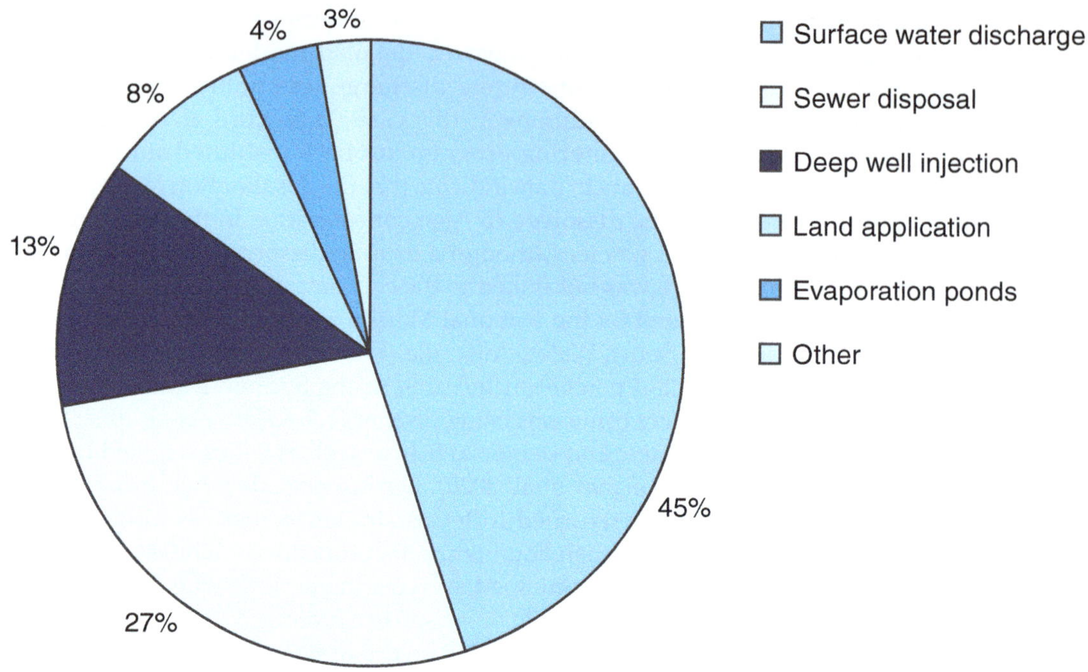


*Adapted from* US Bureau of Reclamation 2018

**Figure 4-1 Current concentrate management practices**

## 4.5 CONCENTRATE DISPOSAL ALTERNATIVES

Numerous concentrate alternatives are utilized for both brackish and seawater desalination facilities. According to the 2018 report on concentrate treatment prepared by the US Bureau of Reclamation (2018), the five most commonly used concentrate management alternatives in the United States are surface water discharge, sewer disposal, deep-well injection, land application, and evaporation ponds (see Figure 4-1). The desalination concentrate management practices shown in the figure have similar frequencies of application worldwide.

The survey included only plants with a capacity larger than 0.05 mgd (0.2 ML/d). Approximately 95 percent of those surveyed were nanofiltration or brackish water plants, with only a small portion representing seawater facilities.

### 4.5.1 Surface Water Discharge

Discharge of seawater desalination facility concentrate through an existing surface water discharge is the most commonly used method worldwide (Table 4-2). Dilution of the concentrate can be achieved by discharging into a high-energy environment or predischarge dilution using flow augmentation. Flow augmentation requires the intake of additional source water for the specific purpose of diluting the concentrate. A surface water discharge using flow augmentation may increase the intake mortality but reduces the turbulence-induced shearing mortality caused by diffuser discharges. Additional cost savings from not building the offshore diffuser result in lower product water costs. In California, only facilities using a surface water intake that were at least 80 percent constructed in 2015 when the California Ocean Plan was amended to regulate seawater desalination are allowed to consider flow augmentation for concentrate dilution if it can be demonstrated that flow augmentation results in less marine life impact than offshore diffusers would cause through turbulence-induced shearing mortality. The

Carlsbad desalination plant qualified to use a surface water discharge incorporating flow augmentation as a potential concentrate dilution in California. It was permitted to operate with this discharge configuration in 2019 by the San Diego Regional Water Quality Control Board with a special NPDES-permit provision to conduct a new focused study to measure and compare flow augmentation impacts versus those likely to be caused by an offshore diffuser. Facilities using SSI may consider the use of flow augmentation without the need to compare impacts with a diffuser.

### 4.5.2 Direct Discharge Through New Ocean Outfall

Discharge of seawater desalination facility concentrate through a new ocean outfall is widely used for projects of all sizes. More than 90 percent of the large seawater desalination facilities currently in operation dispose of concentrate through a new ocean outfall specifically designed and built for that purpose. Examples of large RO seawater desalination facilities with ocean outfalls for concentrate discharge are the 86-mgd (325-ML/d) plant in Ashkelon, Israel; the 36-mgd (136 ML/d) Tuas Seawater Desalination Plant in Singapore; the 14-mgd (53 ML/d) Larnaka Desalination Facility in Cyprus; and most of the large plants in Spain, Australia, and Middle Eastern countries.

The main purpose of every ocean outfall is to dispose of the plant concentrate in an environmentally safe manner, minimizing the size of the zone of discharge in which the salinity is elevated outside of the typical range of tolerance of the marine organisms inhabiting the discharge area. The two key options available to accelerate concentrate mixing from an ocean outfall discharge are either to rely on the naturally occurring mixing capacity of the tidal (surf) zone or to discharge the concentrate beyond the tidal zone and to install diffusers at the end of the discharge outfall to improve mixing. Although the tidal zone carries a significant amount of turbulent energy and usually provides much better mixing than the end-of-pipe type of diffuser outfall system, this zone has a limited capacity to transport and dissipate the saline discharge load into the open ocean. If the mass of the saline discharge exceeds the threshold of the tidal zone's salinity load transport capacity, the excess salinity will begin to accumulate in the tidal zone and could ultimately result in a long-term salinity increase in this zone beyond the level of tolerance of the aquatic life. Therefore, the tidal zone is usually a suitable location for discharge only when it has adequate capacity to receive, mix, and transport the salinity discharge from a desalination facility into the open ocean.

This salinity threshold mixing–transport capacity of the tidal zone can be determined using hydrodynamic modeling. If the desalination facility TDS discharge load is lower than the tidal zone threshold mixing–transport capacity, concentrate disposal to this zone is preferable and is much more cost effective than the use of a long open outfall equipped with a diffuser system. An example of onshore discharge in the tidal zone is that of the Ashkelon seawater desalination facility.

For small plants (i.e., plants with production capacity of 0.1 mgd [0.4 ML/d] or less), the ocean outfall is usually constructed as an open-ended pipe that extends to 300 ft (91 m) into the tidal zone of the ocean. This type of discharge usually relies on the mixing turbulence of the tidal zone to dissipate the concentrate and to quickly bring the discharge salinity to ambient conditions.

The majority of ocean outfalls for large seawater desalination facilities extend beyond the tidal zone. Large offshore ocean outfalls are usually equipped with diffusers in order to provide the mixing necessary to prevent the heavy saline discharge plume from accumulating at the ocean bottom in the immediate vicinity of the discharge. The length, size, and configuration of the outfall and diffuser structures for a large desalination facility are typically determined based on hydrodynamic or physical modeling for the site-specific conditions of the discharge location.

The key challenges associated with selecting the most appropriate location for the desalination facility's ocean outfall discharge are

- finding an area void of endangered species and stressed marine habitats,
- identifying a location with strong ocean currents that allow quick and effective dissipation of the concentrate discharge,
- avoiding areas with busy marine vessel traffic that could damage the outfall facility and change mixing patterns, and
- selecting a discharge location in relatively shallow waters that are close to the shoreline to minimize outfall construction expenditures.

Some key advantages of constructing a new ocean outfall are that this type of concentrate disposal option can accommodate practically any size of seawater desalination facility and provides for more freedom in selecting a plant location as compared to the other two open-ocean disposal approaches where existing outfalls are used.

Key disadvantages of this discharge approach are that it is usually very costly and that its implementation requires extensive environmental and engineering studies. Depending on the site-specific conditions, the costs for a new ocean outfall could be significant and may range from 5–30 percent of the total desalination facility construction expenditures. The higher end of this range tends to apply to large desalination facilities (i.e., facilities of freshwater production capacity of 10 mgd [38 ML/d] or more) with long outfall pipelines, where the construction of a new concentrate outfall may represent more than 20 percent of the total facility cost.

### 4.5.3 Discharge Through Existing Wastewater Treatment Plant Outfall

Discharging concentrate through an existing wastewater treatment plant outfall would generally involve direct disposal through the plant outfall. While it would also be possible to discharge into a nearby wastewater collection system, thereby indirectly utilizing the wastewater outfall, this approach would be expected to have a considerable impact on the wastewater treatment facility operation while raising concerns about corrosion within the collection system. The key feature of the combined discharge method is the accelerated mixing due to the blending of concentrate that is heavier than ocean water with the lighter wastewater discharge. Depending on the volume of the concentrate and on how well the two waste streams are mixed before the point of discharge, the blending may reduce the size of the wastewater discharge plume and dilute some of its constituents. This co-discharge with the lighter-than-seawater wastewater effluent would also accelerate the dissipation of the saline plume by floating this plume upward and expanding the volume of the ocean water with which it mixes.

Direct discharge through an existing wastewater treatment plant outfall is the preferred method for concentrate discharge in California, pursuant to the Ocean Plan Amendment. This method is currently utilized for the recently recommissioned Charles E. Meyer Desalination Plant in Santa Barbara, a 3,125–acre ft/year (2.8-mgd or 10.6-ML/d) facility. This method is also the proposed solution for the Doheny Ocean Desalination Facility (a 5-mgd [19-ML/d] SSI project) and California American Water's Monterey Peninsula Water Supply Project (a 6.9-mgd [26.1-ML/d] SSI project).

One key consideration related to the use of an existing wastewater treatment plant outfall for direct seawater desalination facility concentrate discharge is the availability and cost of wastewater outfall capacity. For this concentrate disposal option to be feasible, an existing wastewater treatment plant in the vicinity of the desalination facility must exist, and this plant must have available outfall discharge capacity. Additionally, the fees associated with the use of the wastewater treatment plant outfall must be reasonable.

Another key consideration is the potential for WET of the blended discharge that may result from ion imbalance of the blend of the two waste streams. The wastewater utility involved must be comfortable with handling any environmental impacts from the blended discharge or potential separation of liability between the desalination facility and the wastewater treatment plant.

In an example of blending increasing WET, bioassay tests completed on blends of desalination facility concentrate and wastewater effluent from the El Estero Wastewater Treatment Plant in Santa Barbara, Calif., indicate that the blend could exhibit toxicity on fertilized sea urchin *(Strongylocentrotus purpuratus)* eggs. Parallel tests on desalination facility concentrate diluted to similar TDS concentration with seawater rather than wastewater effluent did not show such toxicity effects on sea urchins. Similarly, long-term exposure of red sea urchins to the blend of concentrate from the Carlsbad seawater desalination demonstration plant and ambient seawater discharged by the adjacent Encina Power Plant confirms that sea urchins can survive elevated salinity conditions when the discharge is void of wastewater.

The most likely factor causing the toxicity effect on the sensitive marine species is the difference in ratios between the major ions (calcium, magnesium, sodium, chlorine, and sulfate) and TDS that occur in the wastewater effluent–concentrate blend as compared to the concentrate–seawater blend and the ambient ocean water. Because the SWRO membranes reject all key seawater ions at approximately the same level, the ratios between the concentrations of the individual key ions that contribute to the seawater salinity (calcium, magnesium, sodium, chlorine, and sulfate) and the TDS in the concentrate are approximately the same as the ratios in the ambient seawater. Depending on the individual source of wastewater, the ionic ratio for the wastewater likely varies somewhat from ambient seawater, which causes the toxicity effect.

Another important issue to be considered when an existing wastewater discharge with diffusers is used for co-disposal of wastewater and seawater concentrate is the fact that the buoyancy of the mix will be reduced, and the wastewater discharge may no longer provide adequate mixing unless the diffusers are reconfigured. The need for diffuser reconfiguration to accommodate mixed wastewater effluent/concentrate discharge can be established by hydrodynamic modeling.

Diffuser reconfiguration may involve closing (capping) some of the existing diffusers, modifying the discharge diffuser structure, or pumping the mix of concentrate and wastewater influent in order to increase the kinetic energy of the discharge that is available for mixing with the ambient seawater.

Use of existing wastewater treatment plant outfalls for concentrate discharge also has some key advantages, including avoiding costs and environmental impacts associated with the construction of a new outfall for the seawater desalination facility. Mixing of the positively buoyant wastewater discharge with the negatively buoyant concentrate promotes accelerated dissipation of both the wastewater plume and the concentrate. In addition, concentrate often contains metals, organics, and pathogens at significantly lower levels than the wastewater discharge, which reduces the overall waste discharge concentration of the mix.

Although the use of existing wastewater treatment plant outfalls or concentrate discharge to the sanitary sewer system may seem attractive for its simplicity and low construction costs, this disposal method has a number of limitations. Because of the potential toxicity of the concentrate–wastewater effluent blend, the direct discharge of the seawater concentrate through existing wastewater discharge outfalls may be limited. Similarly, indirect discharge of the concentrate through the wastewater collection system may be severely constrained or practically impossible, especially if the wastewater plant effluent is reused for irrigation (Voutchkov and Kaiser 2020).

### 4.5.4 Colocated Discharge Through Existing Operating Power Plant Outfall

It is important to consider the clear distinction between colocated operations of a seawater desalination plant and a coastal once-through cooling power plant and a seawater desalination plant repurposing existing infrastructure left by a decommissioned power plant. The key feature of the colocation concept is the direct connection of the membrane desalination facility intake and discharge facilities to the discharge outfall of an adjacently located coastal power-generation plant. This approach allows the use of the power plant cooling water as both the source water for the seawater desalination facility and as a blending water to reduce the salinity of the desalination facility concentrate before the discharge to the ocean. Colocation of large-scale desalination with a power station was first utilized in the United States for the Tampa Bay Seawater Desalination Project and since then has been considered for numerous plants in the United States and worldwide. The intake and discharge of the Tampa Bay plant are connected directly to the cooling water discharge outfalls of the Tampa Electric Big Bend Power Station.

An example of a colocation configuration where the power plant discharge is used only for dilution of the concentrate and not as a source of feedwater is the 32-mgd (121 ML/d) Carboneras seawater desalination facility in Almería, Spain, which currently is one of the largest SWRO plants in Europe. The concentrate is discharged to the cooling water canal of a nearby coastal power-generation plant and diluted to environmentally safe levels before returning to the sea. The Carboneras facility has a separate open intake independent from the power plant.

In California, the use of once-through cooling by power plants has been largely curtailed with the 2010 passage of the statewide Water Quality Control Policy on the Use of Coastal and Estuarine Waters for Power Plant Cooling. Presently, no seawater desalination plants are planned in California to operate colocated with a power plant, although some plan to use existing power plant offshore intake and outfall structures after the power plant is decommissioned and ceases use of once-through cooling. California regulators consider the tradeoff of reusing existing infrastructure with minimal construction impacts versus new construction impacts when evaluating the best intake site location. Such seawater desalination facilities are not considered colocated by the California regulators. Instead, they are classified as standalone seawater desalination facilities that are repurposing the existing infrastructure. Outside of California in the United States, the implementation of Section 316(b) of the US Clean Water Act may similarly impede future colocation operational opportunities for seawater desalination plants and power plants.

Where possible, colocation with an existing once-through cooled coastal power plant yields four key benefits:

- the construction of a separate desalination facility outfall structure is avoided, thereby reducing the overall cost of desalinated water;
- the salinity of the desalination facility discharge is reduced as a result of the mixing and dilution of the membrane concentrate with the power plant discharge that has ambient seawater salinity;
- because a portion of the discharge water is converted into potable water, the power plant thermal discharge load is decreased, which in turn lessens the negative effect of the power plant thermal plume on the aquatic environment; and
- the blending of the desalination facility and the power plant discharges results in accelerated dissipation of both the salinity and the thermal discharges.

The cost of construction of a separate ocean outfall could be significant, and its avoidance would result in a measurable reduction of plant construction expenditures. In addition, the length and configuration of the desalination facility concentrate discharge outfall are closely related to the discharge salinity. Usually, the lower the discharge salinity, the shorter the outfall and the less sophisticated the discharge diffuser configuration needed to achieve environmentally safe concentrate discharge. Blending the desalination facility concentrate with the lower salinity power plant cooling water reduces the overall salinity of the ocean discharge within the range of natural variability of the ocean, thereby completely eliminating the need for complex and costly discharge diffuser structures.

### 4.5.5 Deep Well Injection and Coastal Well Disposal

Deep well injection involves the discharge of desalination facility concentrate into an acceptable confined deep underground aquifer below the freshwater aquifer(s) using a system of disposal wells. The deep-well injection concentrate disposal system also includes a set of monitoring wells to confirm that the concentrate is not migrating to the adjacent aquifers. A variation of this disposal alternative is the injection of concentrate into existing oil and gas fields to aid field recovery.

Deep well injection is frequently used for concentrate disposal from all sizes of brackish water desalination facilities. Beach well disposal is an alternative concentrate disposal practice.

Unlike deep well injection, beach well disposal consists of concentrate discharge into a relatively shallow unconfined coastal aquifer that ultimately conveys the discharge into the open ocean through the ocean bottom. Beach wells are used for small and medium seawater desalination facilities and are not discussed further in this manual due to limited application and success record.

Deep well injection is most widely used for brackish water discharge. This method of concentrate disposal has found very limited application for seawater desalination. Key considerations associated with this concentrate disposal method include the following.

- Deep well injection is limited to site-specific conditions of confined aquifers of large storage capacity that have good soil transmissivity (Schwartz 2000).
- This disposal method is not feasible for areas of elevated seismic activity or near geologic faults that can provide a direct hydraulic connection between the discharge aquifer and a water supply aquifer.
- There is a potential for contamination of groundwater with concentrated pollutants if the discharge aquifer is not adequately separated from the water supply aquifer in the area of discharge.
- Leakage from the wells is possible.
- There is a potential for scaling and decrease in well discharge capacity over time.
- A backup concentrate disposal method is required for periods of time when the injection wells are tested and maintained.
- High well construction and monitoring costs are associated with deep well injection.

The cities of Marina and Sand City, Calif., have used shallow coastal well discharge of seawater desalination facility concentrate. The discharges inject concentrate with salinity of 30,000–43,000 mg/L of TDS into shallow dune sand aquifers via a conventional well for Marina and a horizontal well for Sand City. The injected concentrate blends with groundwater and ultimately diffuses into the turbulent surf zone. At present, the concentrate from the Marina SWRO plant is no longer discharged through the beach well because of severe scaling problems. The 0.6-mgd (2.3-ML/d) Sand City desalination facility began operation in 2009 and continues to discharge through its subsurface discharge well. This plant uses intake wells to collect source seawater (typical salinity of 17,000–28,000 mg/L) for the production of freshwater and collects additional seawater to dilute the desalination facility concentrate to ambient seawater salinity levels as needed (Wetterau et al. 2009).

A study in Spain suggests that actual dilution of the concentrate from a beach-discharge outfall may be lower than normally accepted (Fernández-Torquemada et al. 2005). In this case, elevated salinity was reported in deep localities several miles from the discharge point. Similarly, a modeling study from Oman suggests that continuously discharging concentrate directly on the shoreline may result in increased salinity along the coastline (Purnama et al. 2003).

### 4.5.6 Beneficial Uses of Concentrate

Concentrate from seawater desalination facilities contains large quantities of minerals that may have commercial value when extracted. The most valuable minerals from the seawater are magnesium, calcium, sodium, chloride, and bromide. Magnesium compounds in seawater have agricultural, nutritional, chemical, construction, and industrial applications. Calcium sulphate (gypsum) is used as construction material for wallboard, plaster, building cement, and road building and repair. Sodium chloride can be used for production of chlorine and caustic soda, highway de-icing, and food products. Technologies for beneficial recovery of minerals from concentrate can be used for management of concentrate from both inland brackish water desalination facilities and coastal seawater desalination facilities. These technologies have the potential to decrease the volume and cost of transporting concentrate as well but are not commercially viable on a large scale and to date are only used in a limited capacity.

The existing salt recovery technologies extract salts by fractional crystallization or precipitation. Crystallization of a given salt can be achieved by concentrate evaporation or temperature control. Fractional precipitation is attained by adding a chemical precipitating agent to selectively remove a target mineral from the concentrate solution. For example, there is a commercially available technology for extraction of magnesium and calcium salts from concentrate and for production of structural materials from these salts.

The sustainable existence of closed systems like our planet has to rely on an efficient circular path when using resources like energy and water. A circular economy is the only path forward toward worldwide sustainable economic growth. For example, applying the circular economy model, brine generated from desalination plants can be used as a source of valuable minerals, such as calcium, magnesium, and sodium chloride. Rare-earth elements can also be extracted from brine, including lithium, strontium, thorium, and rubidium.

Recent stresses in the global market of rare-earth elements have brought the availability and supply of rare metals to the forefront of the sustainability debate and research agenda. These metals are used to fabricate critical components of numerous products, including airplanes, automobiles, smart phones, and biomedical devices. There is a growing realization that the development and deployment of clean energy technologies and sustainable products, processes, and manufacturing industries of the 21st century will also require large amounts of rare metals and valuable elements including platinum group metals (ruthenium, rhodium, palladium, osmium, iridium, and platinum), lithium, copper, cobalt, silver, and gold.

The latest technology trends show that magnesium is replacing aluminum in the car, computer, and cell phone industries because it is over 30 percent lighter. While the world's mining sources of magnesium are fairly limited, seawater brine contains very large quantities of magnesium that could be recovered by concentration of desalination brine followed by selective extraction by adsorption.

Over the last several years, the desalination industry has developed a number of brine concentration and mineral extraction technologies that enable the manufacture of commercially valuable products from the brine. Extracting minerals from seawater is a more environmentally friendly enterprise than terrestrial mining. Moreover, seawater extraction will not require freshwater for processing nor create volumes of contaminated water or tailings for disposal. In addition, these new brine concentration technologies enable dramatic reduction or complete elimination of brine discharge to the sea.

This example shows that transitioning to a circular economy can achieve far more than reducing the negative impacts of the linear economy. Rather, it represents a systemic shift that builds long-term resilience, generates business and economic opportunities, and provides environmental and societal benefits. Revenue from sales of products extracted from the brine like sodium chloride can offset the costs for freshwater production.

Beneficial reuse of brine could also be the key to solving the energy sustainability challenges of desalination. As a further example, a type of next-generation nuclear power plant will use thorium and rubidium as a power source instead of uranium, and these elements are readily extractable from the brine.

## 4.6 MANAGEMENT OF DESALINATION PLANT PHYSICAL RESIDUALS

The key residuals generated at a desalination facility, apart from the concentrate discussed previously, include pretreatment process waste streams and spent membrane cleaning solutions. The most commonly used methods for management of these residuals are summarized in Table 4-4.

### 4.6.1 Management of Pretreatment Process Residuals

Table 4-4 presents types of residuals that may be produced in the pretreatment process before applying membrane desalination. The amount of residual material produced is primarily a function of the feedwater quality relative to the constituents that must be removed before the membrane desalination process. Seawater from open ocean intakes contains significant levels of suspended solids. These solids must be removed before RO, either in a backwash stream or as sludge. Other than the concentrate stream, these solids create the most significant residual stream from a desalination facility.

Spent filter backwash water is a waste stream produced by the membrane plant's pretreatment filtration system. Depending on the type of pretreatment system used (granular or membrane filters), the spent filter backwash water will vary in quantity and quality. In general, the membrane pretreatment systems produce 1.5–2 times larger volume of spent filter backwash water than the granular media filters. However, compared to microfiltration (MF) or ultrafiltration (UF) membrane pretreatment filters, granular media filters typically require larger dosages of coagulant for pretreatment. Depending on the source water quality, membrane use may allow successful pretreatment without coagulant addition. Spent pretreatment filter backwash water may include filter aids and coagulants.

Discharge to the ocean, along with plant concentrate without treatment, is one of the most widely practiced disposal methods for spent filter backwash water internationally. This is typically the lowest cost disposal method because it does not involve any treatment before disposal. However, the practice is less common in more environmentally sensitive areas, such as Australia and the United States.

Onsite treatment before surface water discharge or recycle upstream of the filtration system is another option. The filter backwash water must be treated when its direct discharge does not meet water quality requirements. Typically, the most widely used granular media backwash treatment method is gravity settling in lamella plate sedimentation tanks.

Spent washwater from membrane pretreatment systems is less likely to contain coagulants than media filtration backwash, making it more amenable to discharge with the RO concentrate. Where membrane filtration backwash is recovered, it will employ similar processes as are used with media filters. The solid residuals (sludge) retained in the sedimentation basin are generally dewatered onsite in a designated solids handling facility.

**Table 4-4 Residuals from seawater desalination processes**

| Residual | Source or Cause | Most Common Disposal Methods |
|---|---|---|
| Filter and screening backwash solids/ sludge | Suspended solids in the feedwater<br>May contain coagulants and/or filter aid polymers | Settling/thickening, dewatering by belt filter presses, and sludge disposal to sanitary landfill |
| Backwash water | From removal of suspended solids in the feed water | Recycling to the pretreatment filtration system for reuse or disposal with concentrate |
| Cleaning solutions | Cleaning of filtration membranes (MF/UF) and RO membranes | Blending with concentrate and disposal to surface water body<br>Disposal to sanitary sewer |
| Spent media (sand, anthracite, and/or garnet) | From the removal of suspended solids in the source water | Disposal to sanitary landfill |
| Cartridge filters, polypropylene | Final fine filtration before RO, periodic replacement | Disposal to sanitary landfill |
| MF/UF pretreatment membranes, polymeric material (polypropylene, polysulphone, polyvinylidene-fluoride, cellulose acetate | Membrane replacement for MF/UF pretreatment systems | Disposal to sanitary landfill<br>Regeneration by membrane manufacturer and reuse |
| RO membranes (polyamide thin film composite, cellulose acetate) | Membrane replacements | Disposal to sanitary landfill<br>Regeneration by membrane manufacturer and reuse |

*Source:* Voutchkov 2017
MF/UF — microfiltration/ultrafiltration, RO — reverse osmosis

**Table 4-5 Comparison of waste streams from granular media and membrane pretreatment**

| Waste Stream | Granular Media Filtration (% of Feed Volume) | Membrane Filtration (% of Feed Volume) |
|---|---|---|
| Intake bar screens washwater | 0.1–0.2 | 0.1–0.2 |
| Microscreen washwater | None (not needed) | 0.1–0.5 |
| Spent filter backwash water (reject) | 2.0–6.0 | 5.0–10.0 |
| Chemically enhanced backwash | None (not needed) | 0.2–0.4 |
| Spent membrane cleaning chemicals | None (not needed) | 0.03–0.05 |
| Total waste (% of feed volume) | 2.1–6.2 | 5.4–11.1 |

*Source:* Voutchkov 2017

Conventional granular media filters and membrane pretreatment systems differ significantly by the type, quality, and quantity of the residuals generated during the filtration process (see Table 4-5).

Typically, granular media filtration systems generate only one large liquid waste stream, containing both spent filter backwash and filter-to-waste flows. The volume of this stream in a well-designed plant is 2–6 percent of the total plant intake seawater volume. In addition to the particulate solids and colloids that are contained in the source seawater, this waste stream will also contain coagulants (typically iron salt) and may have flocculants (polymer).

Membrane pretreatment systems generate two primary liquid residual streams: spent membrane backwash water (reject) and membrane-cleaning solution from daily or periodic chemically enhanced backwash (CEB). The volume of the spent membrane filter backwash water is typically 5–10 percent of the plant intake source volume: i.e., approximately 1.5–2 times larger than the spent filter backwash water volume of granular media pretreatment systems.

The difference in total liquid residual volume generated by membrane pretreatment systems is even larger, considering that the microscreens needed to protect the membrane pretreatment filters will be a source of an additional waste discharge from their intermittent cleaning. While conventional traveling fine bar screens use 0.1–0.2 percent of the intake source water for cleaning, the microscreens generate waste screen-wash volumes that equal a slightly higher 0.1–0.5 percent of the intake flow. The larger waste stream volume of the membrane pretreatment system would require a proportionally larger intake seawater volume, which in turn would result in increased size and construction costs for the desalination facility intake facilities and pump station and in higher O&M costs for source seawater pumping to the pretreatment facilities.

In addition to backwashing, cost-competitive design and operation of membrane pretreatment systems often require daily or periodic CEB using a large dosage of chlorine (typically 20–200 mg/L) and strong base and/or acid over a short period of time. This performance-enhancing CEB adds to the volume of the waste streams generated at the RO membrane plant and to the overall cost of source water pretreatment. The daily volume of waste stream generated during CEB is usually 0.2–0.4 percent of the volume of the intake source seawater.

Another waste stream that is associated only with membrane pretreatment is generated during the periodic chemical cleaning of the pretreatment membranes. Extended off-line chemical cleaning, often referred to as clean-in-place (CIP), is usually needed once every 1–3 months. During CIP, membranes are soaked in a solution of hydrochloric or citric acid, sodium hydroxide, sodium hypochlorite, and/or surfactants. CIP is critical for maintaining steady state membrane performance and productivity, but the cleaning generates an additional waste stream that is 0.03–0.05 percent of the source seawater volume.

One advantage of membrane pretreatment systems is that the waste filter backwash generated by these systems contains fewer source water conditioning chemicals (coagulant and polymer), and therefore it is more environmentally benign compared to the waste filter backwash stream generated by conventional pretreatment facilities. This benefit stems from the fact that the typical coagulant dosage for seawater pretreatment by membrane filtration, if used at all, is 2–3 times lower than that for granular media filtration (Voutchkov 2017).

In many cases, source seawater may not need to be conditioned with a coagulant before membrane pretreatment, and this spent filter backwash could be disposed of along with the SWRO concentrate without further treatment. Conversely, due to the high content of iron, the spent filter backwash from granular media filtration pretreatment would need to be treated by sedimentation and other processes, with the settled solids dewatered and disposed to sanitary landfill; otherwise, the high content of iron salt in the backwash water will cause the desalination facility discharge to have a red color, as has been the case for some seawater outfalls (Voutchkov and Kaiser 2020).

The waste streams generated during the CEB and the CIP membrane cleaning should be pretreated onsite in a neutralization tank before discharge. The additional treatment and disposal costs of the waste membrane cleaning chemicals should be considered when comparing membrane and granular media pretreatment systems.

### 4.6.2 GHG Emissions

GHGs are gases that trap heat in the atmosphere. Some GHGs like carbon dioxide occur naturally and are emitted to the atmosphere through natural processes as well as human activities. Other GHGs (e.g., fluorinated gases) are primarily from anthropogenic activities. The principal GHGs that enter the atmosphere through human activities are carbon dioxide, methane, nitrous oxide, and fluorinated gases.

**Carbon dioxide ($CO_2$).** Anthropogenic carbon dioxide enters the atmosphere through the burning of fossil fuels (oil, natural gas, and coal), solid waste, trees and wood products, and various other chemical reactions (e.g., manufacture of cement). Carbon dioxide is also removed from the atmosphere (or sequestered) when it is absorbed by plants as part of the biological carbon cycle or when it is dissolved into water as carbonic acid. As discussed previously in this chapter, sequestration of carbon dioxide into the ocean through OAE has been proposed as a method for using desalination concentrate to reduce atmospheric GHGs.

**Methane ($CH_4$).** Methane is emitted during the production and transport of coal, natural gas, and oil. Methane emissions are also caused by livestock and other agricultural practices and by the decay of organic waste in municipal solid waste landfills.

**Nitrous oxide ($N_2O$).** Nitrous oxide is emitted during agricultural and industrial activities, as well as during combustion of fossil fuels and solid waste.

**Fluorinated gases.** Hydrofluorocarbons, perfluorocarbons, and sulfur hexafluoride are synthetic, powerful GHGs that are emitted from a variety of industrial processes. These gases are typically emitted in smaller quantities, but because they are potent GHGs, they are sometimes referred to as high global warming potential gases.

Changes in the atmospheric concentrations of these GHGs can alter the balance of energy transfers between the atmosphere, space, land, and the oceans and ultimately result in global and local climate variability and permanent changes (USEPA 2025a). Many elements of human society and the environment are sensitive to climate variability and change. Human health, agriculture, natural ecosystems, coastal areas, and heating and cooling requirements are examples of climate-sensitive systems. The extent of climate change effects and whether these effects prove harmful or beneficial will vary by region, over time, and with the ability of different social and environmental systems to adapt to or cope with the change.

Rising average temperatures are already affecting the environment. Some observed changes include shrinking of glaciers, thawing of permafrost, later freezing and earlier breakup of ice on rivers and lakes, lengthening of growing seasons, shifts in plant and animal ranges, and earlier flowering of trees. Global temperatures are expected to continue to rise as human activities continue to add carbon dioxide, methane, nitrous oxide, and other GHGs to the atmosphere. Most of the United States is expected to experience an increase in average temperature as a result of this increase in GHG emissions (USEPA 2025b).

According to USEPA GHG emission inventories, the primary greenhouse gas emitted by human activities in the United States in 2022 was carbon dioxide, representing approximately 79.7 percent of total greenhouse gas emissions (USEPA 2024). The largest sources of carbon dioxide and of overall greenhouse gas emissions are fossil-fuel based production of electricity and transportation. Despite the attention of some environmental groups to greenhouse emissions associated with water production, both conventional and membrane water treatment plants are typically not major sources of GHGs (USEPA 2025b). Water-related energy use is estimated to account for 4 percent of the nation's electricity generation (Copeland and Carter 2017).

### 4.6.3 Greenhouse Gas Emission Impacts and Mitigation

In the United States, evaluating and mitigating the impact of GHG emissions and associated energy consumption is one of the key aspects of ocean desalination design and permitting. Minimization and mitigation of GHG emissions is driven both by a desire to reduce potential adverse effects related to global climate change, reduce project energy consumption, and thereby reduce the desalinated water consumer cost.

In Australia, GHG considerations have played major roles in the implementation of seawater desalination facilities, with desalination facilities in Perth, Queensland, Sydney, Adelaide, and Victoria all employing wind power as their primary energy source. In the United States, there are ongoing efforts to develop more energy-efficient desalination processes, including the US Department of Energy's Office of Energy Efficiency and Renewable Energy, which in 2018 awarded US$21 million in matching grants to study solar thermal desalination (US DOE 2018). Other programs led by the US Bureau of Reclamation and the Water Research Foundation continue to explore improved energy efficiency, including incorporation of renewable energy into desalination projects.

California has particularly strict regulations to reduce and mitigate GHG emissions, as required through California's CEQA process, in addition to requirements set by various regulatory agencies, including local municipality Climate Action Plans, California Coastal Commission, and the state Lands Commission. Specific regulations, policies, and political considerations may vary widely among federal, state, and local agencies with oversight over desalination issues (as is the case with other environmental factors). This makes it more difficult to develop a comprehensive impact assessment and mitigation program, which is often referred to as a GHG mitigation plan or an energy minimization and GHG reduction plan. Although the specific GHG significance threshold may vary between agencies, desalination proponents have addressed GHGs (at least in California) through one of three approaches: reduce the project GHG emissions below a specified significance threshold; reduce the project GHG emissions such that they are equivalent to the GHG emissions that would occur under the current water supply portfolio (mitigate only the incremental additional GHG emissions, referred to as "carbon neutral"); or reduce the project's net GHG emissions to zero without deducting for the comparative emissions of the current water supply portfolio, referred to as "carbon zero." The ocean desalination facility operating in Carlsbad, Calif., was successfully permitted on the basis of a "carbon neutral" GHG mitigation plan (Poseidon Water 2008).

The key step in GHG management is the development of a GHG reduction plan that defines the carbon footprint of desalination facility operations and identifies a portfolio of alternative technologies and measures to achieve carbon footprint project neutrality: from the use of state-of-the-art energy reduction technologies to the implementation of renewable energy projects and carbon dioxide sequestration initiatives including onsite carbon dioxide use, reforestation, and coastal wetland restoration.

An example of the key steps and approaches for the development of an energy minimization and GHG reduction plan is presented in a case study for the Carlsbad seawater desalination facility.

The following are key components of a GHG reduction plan (or "climate action plan").

**Assessing project gross carbon footprint.** The carbon footprint of the seawater desalination facility is the amount of GHG that would be released into the air from the power-generation sources that will supply electricity for the plant. Usually, a carbon footprint is measured in pounds (lb) or metric tons of carbon dioxide emitted per year. The total plant carbon footprint is dependent on two key factors: how much electricity is used by the desalination facility and what sources (fossil fuels, wind, sunlight, etc.) are used to generate the electricity supplied to the plant. Both of these factors could be variable over time, and therefore, the GHG reduction plan has to have the flexibility to incorporate such changes.

The Carlsbad facility operates continuously, 24 h a day/365 days per year, producing an average annual drinking water flow of 50 mgd (189 ML/d). When the plant was originally conceived, the total baseline power use for this plant was projected at 31.3 megawatts (MW) or 4.9 MW•h/acre-ft of drinking water. This power use incorporates both production of fresh drinking water and conveyance and delivery of this water to the distribution systems of the individual utilities and municipalities served by the plant. The total annual energy consumption for the Carlsbad project baseline design is 274,400 MW•h/year. This energy use is determined for an annual average plant production capacity of 50 mgd (189 ML/d). As actual production capacity may vary from year to year, so would the total energy use.

To determine the carbon footprint from the desalination facility annual energy use, the demand is multiplied by the electric grid emission factor, which is the amount of GHGs emitted during the production of unit electricity consumed from the power transmission and distribution system.

The value of the emission factor is specific to the supplier of electricity for the project. San Diego Gas and Electric (SDG&E) supplies electricity to the Carlsbad facility. Similar to other power suppliers in California, SDG&E determines their emission factor based on a standard protocol developed by the California Climate Action Registry (CCAR). CCAR was created by California Legislature (SB 1771) in 2001 as a nonprofit voluntary registry for GHG emissions and is the authority in California that sets the rules by which GHG emissions are determined and accounted for.

Based on information provided in SDG&E's emissions report (CCAR 2008), the SDG&E emission factor is 780.79 lb of carbon dioxide per MW•h of delivered electricity. At 274,400 MW•h/year of energy use and 780.79 lb carbon dioxide/MW•h, the total carbon footprint for the Carlsbad project is calculated at 214.2 million lb of carbon dioxide/year (97,165 metric tons carbon dioxide/year). This carbon footprint is reflective of the latest energy-efficient design of the desalination facility.

The carbon footprint of projects should consider design and site-specific factors. For example, for SSI, some agencies may require consideration of the carbon dioxide "degassing" that may occur when bringing cooler subsurface water under pressure up to the ground surface. GHG emissions attributable to carbon dioxide offgassing are typically a nominal portion of the project's total GHG emissions.

**Reduction due to high-efficiency design.** All existing seawater desalination projects in the US, including the Tampa Bay plant, which began commercial operation on January 25, 2008, are using standard energy recovery equipment. However, utilizing state-of-the-art features can help significantly improve future desalination project's energy efficiency.

For example, one feature of the Carlsbad facility uses a state-of-the-art pressure exchanger–based energy recovery system that allows recovery and reuse of 33.9 percent of the energy associated with the RO process. A significant portion of the energy applied in the RO process is retained in the concentrated stream. These energy exchangers recover and reuse approximately 45 percent of the energy used by the RO process. (Note: The "45 percent energy recovery and reuse" refers to the gross energy recovery potential, while the "33.9 percent energy recovery and reuse" refers to the actual energy savings associated with the energy recovery system.)

In addition to the state-of-the-art pressure exchanger system, the high-energy efficiency design incorporates premium efficiency motors and VFDs on desalination plant pumps that have motors of 500 hp or more. The total desalination plant energy use under the high-energy efficiency design is 28.1 aMW, which corresponds to unit power use of 13.488 kW•h/kgal (3.56 kW•h/m$^3$). The main energy savings result from the use of pressure exchangers instead of Pelton wheels for energy recovery. The pressure exchangers are projected to yield 2,650 hp (2.0 aMW) of power savings, which is a 6.3-percent reduction of the total power use of 31.3 aMW. Converted into unit power savings, the energy reduction of 2.0 aMW corresponds to 0.95 kW•h/kgal (0.25 kW•h/m$^3$). The installation of premium efficiency motors and VFDs on large pumps would result in additional 1.2 aMW (4 percent) of power savings (Poseidon Water 2008).

**Offsetting carbon footprint by reduced water imports.** In many parts of the world like Spain, Israel, Singapore, Australia, and California, seawater desalination facilities are built to replace in- or out-of-state water transfers/supplies. Long-distance water transfers are often energy intensive, and the carbon footprint of such water supply alternatives may be comparable to that of desalination facility of similar capacity. Offsetting the carbon footprint of such long-distance water transfers by building local desalination facilities can be counted as a carbon footprint–reduction measure for the desalination facility.

For example, San Diego County imported water supply dropped from approximately 91 percent in 1991 down to just 11 percent in 2020 through a variety of water supply portfolio diversification programs, including developing the Carlsbad plant, which represents 10 percent of the county's water supply needs (City of Carlsbad, California 2025). The facility receives the majority of its water from two sources: the Sacramento Bay/San Joaquin River Delta, traditionally known as the Bay-Delta, and the Colorado River. The imported Bay-Delta water is withdrawn from the source and conveyed via a complex system of intakes, dams, reservoirs, aqueducts, and pump stations and treated in conventional water treatment plants before its introduction to the water distribution system.

Over the past decade, the availability of imported water from the California State Water Project has been in a steady decline due to prolonged drought, climate change patterns, and environmental and population growth pressures. One of the key reasons for the development of the Carlsbad facility was to decrease reliance on imported water, as described above. Because of these import offsets, this project also offsets the electricity consumption and the GHG emissions associated with pumping, treatment, and distribution of this imported water. The annual energy use for importing 50 mgd (189 ML/d) of state water project water was estimated in 2008 at 190,700 MW•h/year. At 780.79 lb (354.16 kg) carbon dioxide /MW•h, the total carbon footprint of the water imports that is offset by desalinated water is therefore roughly 148.8 million lb of carbon dioxide/year (67,506 metric tons carbon dioxide/year).

The desalination facility also provides opportunity for carbon offsets, as the gross carbon footprint of the desalination facility is 97,165 metric tons of carbon dioxide/year, and 67,506 metric tons of carbon dioxide/year (69.5 percent) of these GHG emissions are offset by reduction of 50 mgd (189 ML/d) of water imports to San Diego County. Additionally, 13,731 metric tons carbon dioxide/year (14.1 percent) would be reduced through other measures identified in the energy minimization and GHG reduction plan, such as green building design, onsite solar power generation, and recovery of carbon dioxide. These measures and reductions are shown in Table 4-6. Thus, the Carlsbad desalination facility's net carbon footprint is estimated at 16,228 metric tons carbon dioxide/year. Note that these estimates were developed as part of the project's GHG reduction plan. Annually, the project's actual GHG emissions are calculated, and the corresponding net incremental emissions required to be offset are verified by a third-party verification entity approved by the state.

**Climate action plan for net carbon footprint reduction.** The main purpose of the climate action plan for a given seawater desalination project is to eliminate a plant's net carbon footprint by implementing measures for energy efficient facility design and operations: green building design, use of carbon dioxide for water production, onsite solar power generation, carbon dioxide sequestration by creation of coastal wetlands and reforestation, funding renewable power–generation projects, and acquisition of renewable energy credits. Project carbon neutrality would be achieved by a balanced combination of these measures.

The size and priority of the individual projects included in the climate action plan should be determined based on a life-cycle cost–benefit analysis and overall benefit for the local community. Implementation of energy efficiency measures for water production, green building design, and carbon dioxide sequestration projects in the vicinity of the project site should be given the highest priority.

The project climate action plan is a living document that must be updated periodically in order to reflect the dynamics of development of desalination and green energy generation technologies. The efficiency and cost-effectiveness of various carbon footprint reduction and offset alternatives must also be updated.

**Green building design.** Whenever practical and viable, the desalination facility should be located on a site of little current value or public use. Reclaiming low-value land will reduce project imprint on the environment as compared to using a new undisturbed site. For example, the Carlsbad facility is located on a site once occupied by dilapidated fuel oil tanks. The tanks and their contents were removed and the site reclaimed and reused to construct the desalination facility.

Another approach to reduce a desalination facility's physical imprint on the environment is to minimize desalination facility site footprint. For example, a key green feature of the Carlsbad facility design is its compactness. The desalination facilities are configured as series of structures sharing common walls, roofs, and equipment that will allow significant reduction of its physical footprint. The total area occupied by the desalination facility facilities is less than 7 acres (3 ha), the smallest footprint desalination facility in the world per unit production capacity (7 acres/50 mgd or 3 ha/190 ML/d). For comparison, the 25-mgd (95-ML/d) Tampa Bay facility occupies 8 acres (3.2 ha); the 70-mgd (265-ML/d) Orange County Groundwater Recharge Project, which also uses an RO system, occupies approximately 40 acres (16 ha); and the 86-mgd (325-ML/d) Ashkelon seawater desalination facility, which currently is the largest operational seawater RO facility in the world, occupies 24 acres (9.7 ha). A plant with a smaller physical footprint will also yield a smaller construction-related carbon footprint resulting in lower construction material expenditures and GHG emissions from construction equipment due to a smaller volume of excavation and concrete works. A reduced construction-site footprint also generates fewer dust emissions and requires less water for dust control.

Whenever economically viable and practical, building design should follow the principles of the Leadership in Energy and Environmental Design (LEED) program. This is a program of the US Green Building Council and was developed to promote construction of sustainable buildings that reduce the overall impact of building construction and functions on the environment by sustainable site selection and development, energy efficiency, materials selection, indoor environmental quality, and water savings. Other programs may be used to verify sustainable design principles including GHG emission reduction, such as the Institute for Sustainable Infrastructure's Envision program.

Consistent with the principles of the LEED program, the desalination facility buildings should include features and materials that allow minimizing energy use for lighting, air conditioning, and ventilation. For example, a portion of the walls of the main desalination facility building of the Carlsbad facility is equipped with translucent panels to maximize daylight use and views to the outside. Nonemergency interior lighting is automatically controlled to turn off in unoccupied rooms and facilities. A monitoring system ensures that the ventilation in the individual working areas in the building is maintained at its design minimum requirements. In addition, building design incorporates water-conserving fixtures (lavatory faucets, showers, water closets, urinals, etc.) for plant staff service facilities and for landscape irrigation.

The green desalination facility buildings should incorporate low-emitting materials and thus pose less risk to the natural environment and building's occupants. Low-emitting paints, coatings, adhesives, sealants, and carpet systems should be used on the interior of the buildings whenever possible. The building design team should include professional engineers that have achieved the LEED Accredited Professional designation and are well experienced with the design and construction of green buildings.

**Onsite solar power generation.** Installation of rooftop photovoltaic (PV) systems for solar power–generation as part of the green building design can further reduce a desalination facility's overall electrical demand. This would vary depending on commercial rates that would provide a positive return/capital investment.

**Use of carbon dioxide for water production.** Approximately 2,100 metric tons (2300 tons or 4.6 million lb) of carbon dioxide/year were originally planned to be used at the desalination facility for posttreatment of the freshwater (permeate) produced by the RO system. Carbon dioxide in a gaseous form will be added to the RO permeate in combination with calcium hydroxide or calcium carbonate to form soluble calcium bicarbonate, which adds hardness and alkalinity to the drinking water for distribution system corrosion protection. In this posttreatment process of RO permeate stabilization, gaseous carbon dioxide is sequestered into soluble form of calcium bicarbonate. Because the pH of the drinking water distributed for potable use is in a range of 8.3–8.5 when carbon dioxide is in a soluble bicarbonate form, the carbon dioxide introduced in the RO permeate would remain permanently sequestered in this form and ultimately would be consumed with the drinking water. The plant designer/owner should require the supplier of carbon dioxide to guarantee that the source of this gas is a waste recovery process (i.e., the gas is generated as a waste sidestream that, if not captured and used, will be released into the atmosphere). This requirement is not very difficult to comply with because most of the commercial suppliers of carbon dioxide use gas generated as a waste from other industrial processes (i.e., ethanol plants, breweries, etc.), which is then purified and sold as a commercial product.

**Carbon dioxide sequestration in coastal wetlands.** In addition to the benefit of marine habitat restoration and enhancement, coastal wetlands also act as a "sink" for carbon dioxide. Tidal wetlands are very productive habitats that remove significant amounts of carbon dioxide from the atmosphere, a large portion of which is stored in the wetland soils. While freshwater wetlands also sequester carbon dioxide, they are often a measurable source of methane emissions. For comparison, coastal wetlands and salt marshes release negligible amounts of GHG, and therefore, their carbon sequestration capacity is not measurably reduced by methane production.

For example, as a part of the Carlsbad project, Poseidon Water (Boston, Mass.) is required to develop 66.4 acres (26.9 ha) of new coastal wetlands in San Diego County. These wetlands are designed to create habitat for marine species similar to those found in the Agua Hedionda Lagoon, from which source seawater is collected for the power plant and for desalination facility operations. As restoration and wetland development continue, these areas will be maintained and monitored over the life of the desalination facility operations. Site-specific research is planned to quantify the actual carbon sequestration capacity of the new wetland system proposed development as a part of the Carlsbad project once the wetland project is completed and is fully functional. Typically, it takes 3–5 years for a coastal wetland project to be fully functional and to begin to yield enhanced habitat and GHG sequestration benefits.

**Sequestration through reforestation.** An alternative approach to offset GHG emissions used by the Carlsbad project was the reforestation of land impacted by wildfires in the fall of 2007. This reforestation effort defined in the energy minimization and GHG reduction plan (Poseidon Water 2008) involved planting of 9,000 trees in areas damaged by San Diego Region wildfires, offsetting approximately 245 metric tons (270 tons) of carbon dioxide/year.

**Carbon emission offsets by investing in renewable energy projects.** An alternative approach to offset GHG emissions of a given desalination project is to invest in renewable energy projects located in the service area of the desalination facility. Recent examples in California include purchasing power produced via landfill gas (Monterey Peninsula Water Supply Project), evaluation of fuel cells and natural gas-powered pumps (Doheny Ocean Desalination Project), and pilot testing solar thermal desalination (the WaterFX [now Wacomet] project in the Central Valley).

**Project annual net-zero carbon emission balance.** Table 4-6 summarizes the total and net carbon footprint estimates of the Carlsbad project and quantifies GHG emission reduction and mitigation options, which are planned to be implemented to reduce the plant net carbon emission footprint to zero. Analysis of data presented in Table 4-6 indicates that for this example case study, up to 83 percent of the GHG emissions associated with seawater desalination and drinking water delivery will be reduced by onsite reduction measures, and the remainder will be mitigated by offsite mitigation projects and purchase of renewable energy credits. It should be noted that the contribution of onsite GHG reduction activities is expected to increase over the useful life (i.e., in the next 30 years) of the project because of the following key reasons.

- In the near future, most power suppliers in the United States are planning to significantly increase the percentage of green power sources in their electricity supply portfolio, which in turn will reduce their emission factor and the net desalination facility carbon footprint. California has committed to an electrical supply renewable portfolio standard of 60 percent by 2030 and requires all the state's electricity to come from carbon-free resources by 2045 (California Public Utilities Commission 2025).
- SDG&E is planning to increase significantly the percentage of green power sources in its electricity supply portfolio, which in turn will reduce its emission factor and the project's net indirect GHG emissions.
- Advances in seawater desalination technology are expected to yield further energy savings and carbon footprint reductions. Over the last 20 years, there has been a 50-percent reduction in the energy required for seawater desalination. This trend is projected to continue in the future.

**Table 4-6 Carlsbad desalination project net greenhouse gas (GHG) emission zero balance**

| Source | Total Annual Power Use (MW•h/year) | Total Annual Emissions (metric tons carbon dioxide/year) |
|---|---|---|
| **Part 1: Identification of GHG emissions** | | |
| Project baseline design | 274,400 | 97,165 |
| **Part 2: Onsite and project-related reduction of GHG emissions** | | |
| Reduction due to high-efficiency design | (28,244) | (10,001) |
| Green building design | (300–500) | (106–177) |
| Onsite solar power generation | (0–777) | (0–275) |
| Recovery of carbon dioxide | NA | (2,100) |
| Reducing energy needs for water recycling | (1,950) | (690) |
| Reduced water importation | (190,641) | (67,506) |
| Sequestration in coastal wetlands | NA | (18–188) |
| Subtotal on-site reduction measures | NA | **(80,421–80,937)** |
| | **Net GHG emissions** | **16,422–16,228** |
| **Part 3: Additional offsite reductions of GHG emissions** | | |
| Sequestration through reforestation | NA | (245) |
| Potential renewable energy partnerships | (0–2,260) | (0–800) |
| Subtotal onsite reduction measures | NA | (245–1,045) |
| Offset and renewable energy certificate purchases | NA | **(16,499–15,067)** |
| | **Net GHG emissions** | **0** |

*Source:* Poseidon Water 2008
NOTES: NA – not applicable. Numbers in parentheses indicate reduction.

The costliest carbon footprint reduction options may be the green building design and the installation of a rooftop solar power–generation system. Development of new coastal wetlands is a very promising carbon footprint reduction option, which could be several times less costly than the construction of a solar panel system. Similarly, reforestation could also be a cost-competitive GHG reduction alternative. As compared to green power–generation alternatives (solar and wind power), reforestation and wetland mitigation have added environmental benefits. For example, the new coastal wetlands developed in relation to a seawater desalination project could create habitat for species that are impacted by the intake operations of the desalination facility via I&E of these species on the intake screens.

For GHG emissions that cannot be reduced or mitigated through site design measures or offsite carbon offset approaches, the ocean desalination facility would typically turn to purchasing GHG offsets as part of a comprehensive energy minimization and GHG reduction plan. Table 4-7 summarizes current estimates for purchasing carbon offset credits as one means of mitigating GHG emission impacts.

Despite the fact that the carbon footprint for production of desalinated water is higher than that of production of drinking water from traditional freshwater resources, it is smaller than many other human activities that improve quality of life, such as food refrigeration, heating of water for domestic use, driving a personal vehicle, or flying. The average carbon footprint of producing desalinated water for one person is 0.11 metric tons of carbon dioxide/year, which is only 3.7 percent of the sustainable carbon footprint of one person of 3 metric tons/carbon dioxide/year (IDA Energy and Environment Committee Forum 2019).

Currently, most desalination plants worldwide are supplied by power generated from fossil fuel. However, a number of recent SWRO desalination plants in Australia have implemented wind-driven power-generation projects, which produce as much power as that used by the desalination plants. Over the last 5 years, a number of Middle East and North Africa countries have taken the initiative to develop a robust portfolio of renewable power–generation plants to provide electricity for seawater desalination (World Bank 2019).

Solar and wind power are the most abundant renewable energy sources worldwide at present. Key advantages of solar- as compared to wind-powered desalination plants in the Middle East are the high intensity and reliability of the power source (i.e., solar irradiation) and their relatively lower construction and O&M costs. However, as with wind farms, a key challenge of solar power supply facilities is the need for a large amount

**Table 4-7 Indicative pricing for greenhouse gas (GHG) offsets based on locational hierarchy as of September 2020**

| Location and Project Type | Unit Cost (US$/ metric ton of carbon dioxide reduced) | Comment From Element Markets (GHG Offset Provider) |
|---|---|---|
| In-state offsets (forestry) | 9.00–14.00 | There is a limited market supply for California-based GHG offsets due to compliance program (AB 32), California regulations, and high demand for offsets. |
| United States–based offsets | 2.75–4.25 | Project types include methane combustion from landfills, biomass, and industrial gases. |
| International offsets | 1.00–2.75 | Project types include solar-, wind-, and hydro-energy projects. |

*Adapted from* Anew Climate

* GHG offset price quote based on transactions in 2020 as well as public data on offset availability across the three locational preferences (US state, United States as a whole, and international) and is subject to change due to market demand.

of land for installation of the renewable energy equipment to supply power to SWRO desalination plants.

Outside of the Middle East, wind-generated power has found wider use than solar power for desalination project supply because of its availability as an energy generation source. Even in countries such as Saudi Arabia and the United Arab Emirates, solar power intensity adequate for steady power generation occurs only for 6–7 h per day and for less than 80 percent of the time.

As a rule of thumb, the land area needed for a PV field to power a 1,000 $m^3$/day (0.2 mgd) SWRO plant is 10 ha (24.7 acres), while a wind farm for the same size SWRO plant requires 20 ha. This is approximately 50 times and 100 times higher, respectively, than the land needed to construct the SWRO desalination plant itself. The total capital cost for construction of a solar power plant to supply all the electricity needed for seawater desalination plant at present is typically 60–80 percent of the capital cost of the desalination plant itself. More detailed discussion on the feasibility of linking renewable power and desalination projects is provided elsewhere (World Bank 2012).

Solar power–driven desalination projects under development at present encompass indirect or direct coupling of conventional SWRO, multistage flash distillation (MSF), or multiple effect distillation (MED) desalination plants with either concentrated solar power (CSP)–generation technologies or PV cells (Palenzuela et al. 2014; Blanco et al. 2011). The most promising combinations of solar power and desalination technologies are PVs with RO and MED systems and CSPs with MSF or MED systems (Pinto and Marques 2017; Moser et al. 2013; Shatat et al. 2013; Al-Karaghouli et al. 2009).

Currently, PV-based SWRO solar desalination is the main focal point of research and full-scale desalination project implementation because of the measurable decrease in solar panel costs over the last 5 years. At present, conventional SWRO desalination plants powered through the electric grid still remain economically more competitive than PV-powered RO or CSP-powered MED configurations, as well as to other combinations of desalination technologies and alternative power sources (Moser et al. 2013; ETSAP and IRENA 2012; Fiorenza et al. 2003).

The water production costs presented in Figure 4-2 include both the capital and O&M expenditures for the desalination plant as well as for the renewable power–generation system (wind, solar, geothermal, etc.). These costs are for autonomous, directly coupled desalination plant and renewable energy sources and do not include the use of electricity from the grid. Costs for desalination plants decoupled from the renewable power–generation system could be lower and would depend on the site-specific project conditions (Al-Karaghouli and Kazmerski 2013; World Bank 2012).

Selecting the most suitable renewable energy–driven desalination technology depends on the size of the plant, the source water and product water quality, the availability of access to the electric power grid, and the type of renewable power technology (Ghaffour et al. 2015).

Desalination based on the use of renewable energy sources can provide sustainable long-term production of freshwater and is expected to become economically attractive in the near future as the costs of renewable energy–production technologies continue to decline and the costs of fossil fuel continue to rise. In addition, environmental externalities associated with fossil fuel–based electricity generation (e.g., the need to offset desalination plant carbon footprint) may offset the difference in energy and water production costs (Gude 2016; Karagiannis and Soldatos 2008).

In parallel with the exploration of renewable power alternatives, the world's leading research centers in the United States, Saudi Arabia, and European countries are working on the development of new generation of energy recovery devices (ERDs), high-pressure pumps, and membranes that aim to bring the total energy use of desalination plants to less than 2.45 kW•h/$m^3$ (9.27 kW•h/ 1,000 gal) and the energy demand of the RO desalination system below 1.8 kW•h/$m^3$ (6.8 kW•h/ 1,000 gal).

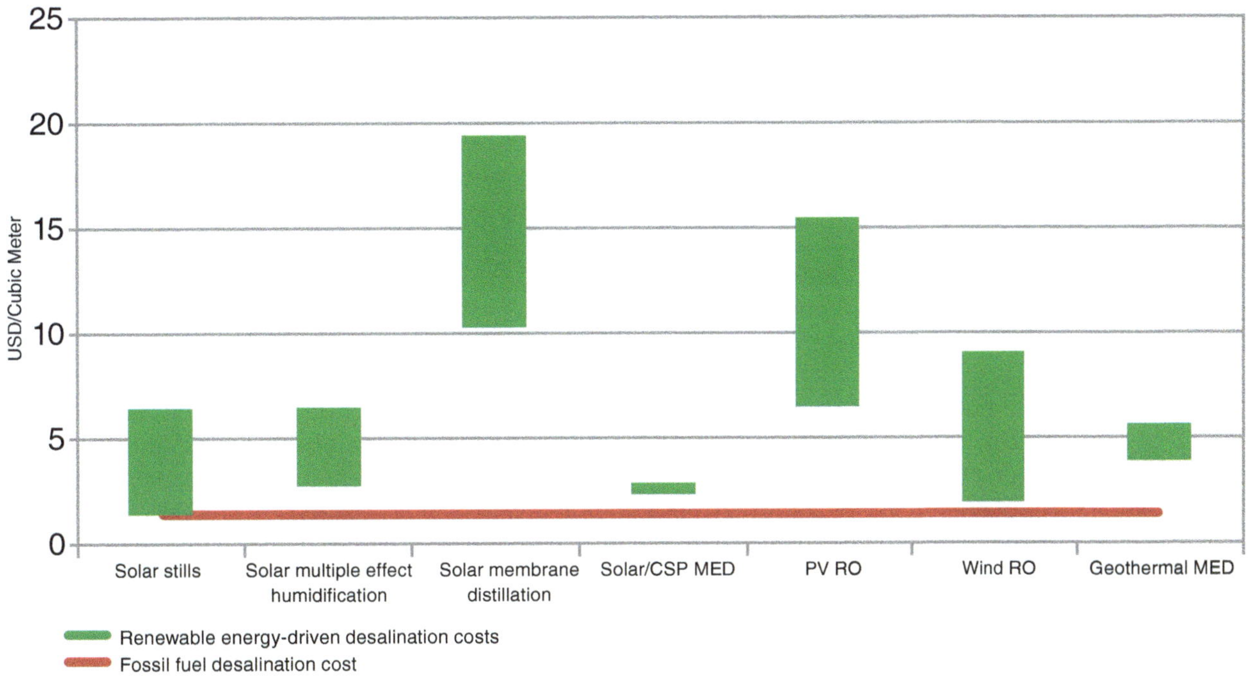


*Source:* Nikolay Voutchkov

CSP MED – concentrated solar power multiple-effect distillation, PV RO – photovoltaic reverse osmosis, wind RO – wind reverse osmosis, geothermal MED – geothermal multiple-effect distillation

**Figure 4-2 Costs of water produced by renewable versus fossil fuel–driven desalination**

These advancements will result in the reduction of the total energy use and carbon footprint of desalination plants by over 30 percent. The new technologies are tailored to fit equally well in both existing desalination plants and future RO facilities.

The new generation ERDs under development at present will be designed to reuse over 98.5 percent of the energy remaining in the brine after membrane separation. Such energy recovery efficiency will exceed significantly the performance of the existing commercially available best-in-class energy recovery technologies and will address some of the flaws of these technologies, such as brine mixing and equipment complexity.

Research and development of the next generation of high-pressure pumps for SWRO systems are projected to have a disruptive impact in reduction in energy use. State-of-the-art high-pressure pumps used for desalination currently have efficiency of 75–83 percent. The new generation pump technologies under the development are targeting efficiency of 95 percent or more. As high-pressure pumps consume 70–75 percent of the total energy in desalination plants, this dramatic improvement of pump efficiency will yield to an unprecedented reduction in desalination plant carbon footprint.

## 4.7 OTHER ENVIRONMENTAL CONSIDERATIONS

Seawater intake, concentrate discharge, and GHG emissions are generally considered the primary issues of focus in the successful siting, design, and regulatory agency approval for an ocean desalination facility. However, depending on site-specific conditions, state and local regulatory requirements, and other factors, numerous other environmental issues could play prominent roles in a project's environmental review and approval process. Some of these factors include the following.

**Local agency regulations.** Local agencies have various regulatory programs that may affect ocean desalination facilities, including restrictions on pumping groundwater, and prohibitions against transporting pumped groundwater between groundwater basins (both of which may be applicable to SSI). For desalination facilities within the coastal zone, local agencies often (usually) have local approval authority over a coastal development permit and may have to demonstrate compliance with the local agency's local coastal program, which requires addressing a wide range of issues from marine resource impacts to aesthetics, coastal access, ESHA, sea-level rise, and others.

**Local stakeholder/Political factors.** Successful development of an ocean desalination facility requires local agency approval and strong stakeholder support. Public outreach and developing strong political support are essential elements of a successful project. Often the success of a project rests on local agency and stakeholder support.

**Growth.** In some cases, ocean desalination may face opposition based upon perceived growth-inducing impacts of introducing a new water supply into a coastal urban environment. In California, particularly where a screened ocean intake is proposed, regulatory agencies scrutinize desalination facility sizing and need in comparison to water supply alternatives perceived as less environmentally harmful.

**Noise, air pollution, and traffic.** Construction and operation of seawater desalination facilities, as is the case with any water facility construction, contribute to all three. Of these, noise is of specific importance because the high-pressure RO feed pumps operate at a very high rotational speed and are usually a significant source of noise pollution. The key sources of noise at the plant are the large high-pressure pumps that feed the RO treatment trains and the interconnected ERDs. These noise sources should be located in the RO building, which would contain the generated noise.

Usually, desalination facilities are equipped with large intake seawater pumps, pretreatment filter transfer pumps, and product water transfer pumps that are often located outdoors. The potential noise mitigation measures for these pumps are as follows.

*Use of centrifugal pumps.* Centrifugal pumps that have relatively low noise levels can be used for these applications as an alternative to higher noise-level piston pumps.

*Use of water-cooled pump motors.* The main source of noise in a centrifugal pump station are the pump motors. Water-cooled motors may be used instead of standard air-cooled motors to reduce noise levels.

*Installation of acoustic enclosures.* Commercially available acoustic enclosures can be installed around the pump motors or the entire pump station to contain and dissipate the noise from the outdoor mechanical equipment.

*Installation of sound curtains.* Industrially sewn sound curtains can be installed around the pump stations using floor-mounted hardware.

*Installation of the pumps and motors in an enclosed building.* If required, the pumps and motors can be installed in an enclosed building designed to attenuate sound sources. Often the noise in the main desalination building, which houses the high-pressure pumps and ERDs, is attenuated by acoustic control panels installed on the walls of the building.

*Special noise design features for SSI in coastal recreational areas.* For SSI facilities in sensitive coastal environs, special design and construction measures may be required: e.g., seasonal construction to avoid recreation area peak seasons, sound curtains where SSI well development is in close proximity to recreational areas or other sensitive receptors, and submersible pumps to minimize long-term operational noise.

## REFERENCES

Al-Karaghouli, A., and L.L. Kazmerski. 2013. "Energy Consumption and Water Production Cost of Conventional and Renewable-energy-powered Desalination Processes." *Renewable Sustainable Energy Review*, 24:343–356.

Al-Karaghouli, A., D. Renne, and L.L. Kazmerski. 2009. "Solar and Wind Opportunities for Water Desalination in the Arab Regions." *Renewable and Sustainable Energy Reviews,* 13:9: 2397–2407.

Avramidi, E., S.C. García Gómez, S. Papaspyrou, V. Louca, D. Xevgenos, and F.C. Küpper. 2022. "Benthic Biodiversity Near Brine Discharge Sites in the Port of Rotterdam." *Water Resources and Industry*, 27:100173.

Blanco, J., D.C. Alarcón-Padilla, G. Zaragoza, P. Palenzuela, E. Guillén, and M. Ibarra. 2011. Solar Energy and Desalination Processes: Status and Development Perspective. Water Desalination Conference in the Arab Countries (ARWADEX), Riyadh, Saudi Arabia.

California Coastal Commission. 2025a. *Laws & Regulations: The Coastal Act*. www.coastal.ca.gov/laws/ (accessed October 27, 2025).

California Coastal Commission. 2025b. *Sea Level Rise: Adopted Policy Guidance*. www.coastal.ca.gov/climate/slrguidance.html (accessed October 27, 2025).

California Ocean Protection Council. 2020. *Strategic Plan to Protect California's Coast and Ocean 2020–2025*. www.opc.ca.gov/webmaster/ftp/pdf/2020-2025-strategic-plan/OPC-2020-2025-Strategic-Plan-FINAL-20200228.pdf (accessed September 10, 2025).

California Public Utilities Commission. 2025. *Renewables Portfolio Standard (RPS) Program.* https://www.cpuc.ca.gov/rps/ (accessed September 10, 2025).

California State Water Resources Control Board. 2019. *Ocean Standards*. https://www.waterboards.ca.gov/water_issues/programs/ocean/ (accessed November 11, 2025).

Carlsbad Desalination Plant. 2019. *New Permit Fosters Sustainable Water Production at Carlsbad Desalination Plant.* www.carlsbaddesal.com/news/new-permit-fosters-sustainable-water-production-at-carlsbad-desalination-plant (accessed September 10, 2025).

Catalanade Perforacions. 2020. *Neodren Seawater Catchment by Horizontal and Directed Drains.* www.catalanadeperforacions.com/en/neodren (accessed October 27, 2025).

CCAR (California Climate Action Registry). 2008. *Annual Emissions Report: San Diego Gas and Electric 2006*. California Climate Action Registry. www.caclimateregistry.org/wp-content/plugins/carrot/carrot_reports/san-diego-gas--electric-2008-ca.pdf (accessed December 9, 2025).

City of Carlsbad, California. 2025. *Seawater Desalination*. https://www.carlsbadca.gov/departments/utilities/water/seawater-desalination (accessed November 11, 2025).

City of Corpus Christi. 2020. *City of Corpus Christi Water System Project Update*. Corpus Christi Seawater Desalination Project.

Clark, G.F., N.A. Knott, B.M. Miller, B.P. Kelaher, M.A. Coleman, S. Ushiama, and E.L. Johnston. 2018. "First Large-Scale Ecological Impact Study of Desalination Outfall Reveals Trade-Offs in Effects of Hypersalinity and Hydrodynamics." *Water Research,* 145:757–768.

Copeland, C., and N.T. Carter. 2017. *Energy-Water Nexus: The Water Sector's Energy Use.* https://sgp.fas.org/crs/misc/R43200.pdf (accessed September 10, 2025).

D.A. Lord & Associates. 2005. *Ecological Assessment of the Effects of Discharge of Seawater Concentrate from the Perth Seawater Desalination Plant on Cockburn Sound*; Report No. 05/028/1. Perth, Australia: D.A. Lord & Associates.

Duff, J., V. LaBate, and A.M. Slugg. 2017. "Prospects and Pitfalls of Desalination Development: Insights From Three States." *Ocean and Coastal Law Journal*, 22:2:article 11.

Einav, R., and F. Lokiec. 2003. "Environmental Aspects of a Desalination Facility in Ashkelon." *Desalination*, 156:1–3:79–85.

ESTAP (Energy Technology Systems Analysis Program) and IRENA (International Renewable Energy Agency). 2012. *Water Desalination Using Renewable Energy*. Technology Brief I12. https://www.irena.org/publications/2012/Mar/Water-Desalination-Using-Renewable-Energy (accessed December 9, 2025).

Fernández-Torquemada, Y., J.L. Sánchez-Lizaso, and J.M. González-Correa. 2005. "Preliminary Results of the Monitoring of the Brine Discharge Produced by the SWRO Desalination Facility of Alicante (SE Spain)." *Desalination*, 182:1–3:395–402.

Fiorenza, G., V.K. Sharma, and G. Braccio, G. 2003. "Techno-economic Evaluation of a Solar Powered Water Desalination Plant." *Energy Conversion and Management*, 44:14:2217–2240.

Florida Department of Environmental Protection, Division of Water Resource Management. 2010. *Desalination in Florida: Technology, Implementation, and Environmental Issues*. https://ufdcimages.uflib.ufl.edu/UF/00/10/03/82/00001/desalination-in-florida-report.pdf (accessed September 12, 2025).

Ghaffour, N., J. Bundschuh, H. Mahmoudi, and M.F.A. Goosen. 2015. "Renewable Energy-driven Desalination Technologies: A Comprehensive Review on Challenges and Potential Applications of Integrated Systems." *Desalination*, 356:94–114 .

Gude, V.G. 2016. "Desalination and Sustainability–An Appraisal and Current Perspective." *Desalination*, 89:87–106.

Hammond, M., N. Blake, P. Hallock-Muller, M. Luther, D. Tomasko, and G. Vargo. 1998. "Effects of Disposal of Seawater Desalination Discharges on Near Shore Benthic Communities." *Report of Southwest Florida Water Management District and University of South Florida*, Tampa, FL.

Hartmann, J., N. Suitner, C. Lim, J. Schneider, L. Marin-Samper, J. Aristegui, P. Renorth, J. Taucher, and U. Riebesell. 2023. "Stability of Alkalinity in Ocean Alkalinity Enhancement (OAE) Approaches – Consequences for Durability of $CO_2$ Storage." Biogeosciences, 20:781–802.

He, J., and M.D. Tyka. 2023. "Limits and $CO_2$ Equilibration of Near-coast Alkalinity Enhancement." *Biogeosciences*, 20:27–43.

IDA Energy and Environment Committee Forum. 2019. "Desalination Impacts on the Environment—Myths and Misconceptions." IDA World Congress, Dubai, UAE: DOI: 10.13140/RG.2.2.22335.05286.

ISTAP (Independent Scientific Technical Advisory Panel). 2015. *Phase 2 Report: Feasibility of Subsurface Intake Designs for the Proposed Poseidon Water Desalination Facility at Huntington Beach, California*. https://documents.coastal.ca.gov/assets/press-releases/huntington-beach-desal/CCC-Poseidon-ISTAP-FINAL-Phase-2-Report-11-09-15.pdf (accessed September 10, 2025).

Karagiannis, I.C., and P.G. Soldatos. 2008. "Water Desalination Cost Literature: Review And Assessment." *Desalination*, 223:1–3:448–456.

Kelaher, B.P., G.F. Clark, E.L. Johnston, and M.A. Coleman. 2020. "Effect of Desalination Discharge on the Abundance and Diversity of Reef Fishes." *Environmental Science & Technology*, 54:2:735–744.

Kelaher, B.P., G.F. Clark, E.L. Johnston, T. Ingleton, N.A. Knott, and M.A. Coleman. 2022. "Desalination Discharge Influences the Composition of Reef Invertebrate and Fish Assemblages." *Environmental Science & Technology*, 56:16:11300–11309.

Mickley, M., and N. Voutchkov. 2016. *Database of Permitting Practices for Seawater Concentrate Disposal*. London: IWA Publishing.

Missimer, T.M, N. Ghaffour, A.H.A. Dehwah, R. Rachman, R.G. Maliva, and G. Amy. 2013. "Subsurface Intakes for Seawater Reverse Osmosis Facilities: Capacity Limitation, Water Quality Improvement, and Economics." *Desalination*, 322:37–51.

Moser, M., F. Trieb, T. Fichter, and J. Kern. 2013. "Renewable Desalination: A Methodology for Cost Comparison." *Desalination and Water Treatment*. 51:4–6:1171–1189.

NSTC (National Science and Technology Council). 2019. *Coordinated Strategic Plan to Advance Desalination for Enhanced Water Security*. https://trumpwhitehouse.archives.gov/wp-content/uploads/2019/03/Coordinated-Strategic-Plan-to-Advance-Desalination-for-Enhanced-Water-Security-2019.pdf (accessed August 29, 2025).

Palenzuela, P., A.S. Hassan, G. Zaragoza, and D.C. Alarcón-Padilla. 2014. "Steady State Model For Multi-Effect Distillation Case Study: Plataforma Solar de Almería MED Pilot Plant." *Desalination*, 337:31–42.

Phillips, B.M., B.S. Anderson, K. Siegler, J.P. Voorhees, S. Katz, L. Jennings, and R.S. Tjeerdema. 2012. *Hyper-Salinity Toxicity Thresholds for Nine California Ocean Plan Toxicity Test Protocols*. www.waterboards.ca.gov/water_issues/programs/ocean/desalination/docs/saltoxfr08012.pdf (accessed September 4, 2025).

Pinto, F.S., and R.C. Marques, 2017. "Desalination Project Economic Feasibility: A Standardization of Cost Determinants." *Renewable and Sustainable Energy Reviews*, 78:904–915.

Poseidon Water. 2008. *CCC Adopted Energy Minimization and Greenhouse Gas Reduction Plan Carlsbad Desalination Project*. www.sdcwa.org/sites/default/files/files/environmental-docs/city-of-carlsbad/2-energy-minimization-greenhouse-gas-reduction-plan-Dec2008.pdf (accessed September 10, 2025).

Purnama, A., H.H. Al-Barwani, and M. Al-Lawatia. 2003. "Modeling Dispersion of Brine Waste Discharges From a Coastal Desalination Facility." *Desalination*, 155:1:41–47.

San Diego County Water Authority. 2020. "New Fish-Friendly Seawater Intake Pumps at Carlsbad Desalination Plant." *Water News Network*. https://www.waternewsnetwork.com/new-fish-friendly-seawater-intake-pumps-at-carlsbad-desalination-plant/ (accessed September 10, 2025).

Schwartz, J. 2000. *Beach Well Intakes for Small Seawater Reverse Osmosis Plants*. Muscat, Sultanate of Oman: The Middle East Desalination Research Center.

Shatat, M., M. Worall, and S. Riffat, S. 2013. "Opportunities for Solar Water Desalination Worldwide: A Review." *Sustainable Cities and Society*, 9:67–80.

State of California Public Utilities Commission. 2019. *Monterey Peninsula Water Supply Project.* https://www.cpuc.ca.gov/Environment/info/esa/mpwsp/PD.html (accessed September 10, 2025).

SWRCB (State Water Resources Control Board). 2015. *California Ocean Plan*. www.waterboards.ca.gov/water_issues/programs/ocean/docs/oceanplan2019.pdf (accessed September 12, 2025).

TWDB (Texas Water Development Board). 2019. *Desalination Seawater.* www.twdb.texas.gov/publications/shells/Desal_Seawater.pdf (accessed November 11, 2025).

US DOE (US Department of Energy). 2018. *Department of Energy Announces $21 Million to Advance Solar Desalination Technologies.* www.energy.gov/articles/department-energy-announces-21-million-advance-solar-desalination-technologies (accessed September 10, 2025).

US Department of the Interior, Bureau of Reclamation. 2018. *Desalination and Water Purification Research and Development Program Report No. 207: Updated and Extended Survey of U.S. Municipal Desalination Plants.* www.usbr.gov/research/dwpr/reportpdfs/report207.pdf (accessed August 29, 2025).

USEPA (US Environmental Protection Agency). 2025a. *Climate Change Indicators: Weather and Climate.* https://www.epa.gov/climate-indicators/weather-climate (accessed September 10, 2025).

USEPA. 2025b. *Overview of Greenhouse Gases*. https://www.epa.gov/ghgemissions/overview-greenhouse-gases (accessed September 10, 2025).

USEPA. 2024. *Inventory of U.S. Greenhouse Gas Emissions and Sinks*. https://www.epa.gov/ghgemissions/inventory-us-greenhouse-gas-emissions-and-sinks (accessed January 15, 2026).

USEPA. 2014. Technical Development Document for the Final Section 316(b) Existing Facilities Rule. www.epa.gov/sites/default/files/2015-04/documents/cooling-water_phase-4_tdd_2014.pdf (accessed September 10, 2025).

Van Senden, D., and B.M. Miller. 2005. *Stratification and Dissolved Oxygen Issues in Cockburn Sound Pertaining to Discharge of Brine From Desalination*. Technical Report 2005/03. Manly Vale, NSW: The University of New South Wales Water Research Laboratory.

Voutchkov, N. 2012. *Desalination Engineering: Planning and Design*. New York: McGraw Hill.

Voutchkov, N. 2013. *Desalination Engineering: Operation and Maintenance*. New York: McGraw Hill.

Voutchkov, N. 2017. *Pretreatment for Reverse Osmosis Desalination*. Amsterdam: Elsevier.

Voutchkov, N., and G.N. Kaiser. 2020. *Management of Concentrate from Desalination Plants*. Amsterdam: Elsevier.

WateReuse Association Desalination Committee. 2012. *Seawater Desalination Power Consumption*. Alexandria, VA: WateReuse Association. https://watereuse.org/wp-content/uploads/2015/10/Power_consumption_white_paper.pdf (accessed September 12, 2025).

World Bank. 2019. *The Role of Desalination in an Increasingly Water-Scarce World*. Washington: World Bank Group.

Wetterau, G., L. Voelz, K. Klinko, R. Siminitch. 2009. "Tapping the Ocean in Monterey County, California: The San City Desalination Facility." 2009 Annual Water Reuse Conference, Seattle, WA.

This page intentionally blank.

Chapter 5

# Cost of Treatment

*Second Edition Authors: Zita L.T. Yu, Rubén Muñoz, Greg Wetterau*
*First Edition Authors: Kurt Kiefer, Tom Pankratz*

## 5.1 INTRODUCTION

During the initial development stages of a seawater desalination project, decision makers use cost models to forecast project costs to determine financial viability and obtain project funding and financing. In developing these cost models, it is important to consider not only the construction capital costs of the desalination plant itself but all of the costs associated with financing, permitting, implementation, and operations and maintenance (O&M) of the facility. It is difficult to generalize some of these costs as they are site specific. In addition, the market has suffered an irregular price disruption in recent years due to the COVID-19 global pandemic and various conflicts impacting supply chains. This chapter therefore includes some discussion on post-2020 price trends and the portions of facility costs that have been most impacted. The chapter is intended to aid in the development of capital, O&M, and life-cycle cost estimates for seawater reverse osmosis (SWRO) desalination projects by providing the latest cost data and industry trends. Key project components that could have significant capital and O&M cost implications—such as source water quality, intakes, treatment, storage and distribution, and disposal of concentrate and other residuals—are discussed in this chapter. A good understanding of the roles of these in the life-cycle cost of the project is an important step in the development of the total project and the final cost of the finished water that will be borne by the customer.

## 5.2 COST OVERVIEW

The costs of SWRO desalination projects can be broadly divided into two categories.

- **Capital costs** include the direct costs for construction of the treatment facility, including land costs, process equipment, auxiliary equipment (transmission piping, storage tanks, pumps, etc.), building costs, material purchase and labor, as well as indirect costs to develop, design, permit, and finance the desalination project.

- **O&M costs** include the cost of labor, energy, chemicals (e.g., cleaning acids, chlorine, antiscalant), auxiliary equipment (e.g., pumps, tanks), pretreatment (e.g., media for filters, dissolved air flotation [DAF], ultrafiltration [UF]/microfiltration [MF]), and other consumables (reverse osmosis [RO] membranes, cartridge filters, etc.) necessary for the O&M of a desalination plant, as well as the costs associated with residuals disposal, project administration, environmental and permit compliance monitoring, and permit maintenance and renewal.

Life-cycle costs that encompass both capital and O&M costs are used to determine the total cost of water and/or to develop life-cycle costs for the project. These costs are discussed in more detail later in this chapter.

## 5.2.1 Capital Costs

Capital costs include two cost components.

- Direct construction costs include the cost of equipment, buildings, pipelines, and other physical facilities that make up the project, including the contractor's general and administrative costs, costs of bonding and insurance, profit, and financing costs. These costs are often referred to as hard capital costs and may represent 60–80 percent of the total capital costs.
- Indirect costs include the cost of engineering, environmental, and siting issues and the legal, administrative, and financial services needed to plan, design, and permit the project. These costs are often referred to as soft costs and may represent 20–40 percent of the total capital costs.

The construction cost of an SWRO desalination project, which accounts for the major portion of the capital costs, can be affected by many factors, including

- source water quality;
- intake type and distance from the plant;
- product water quality goals;
- treatment capacity and production reliability;
- conveyance systems to deliver finished/treated water to distribution networks;
- facilities for processing and/or disposing of residuals (concentrate and pretreatment solids);
- provision for future plant production expansions; and
- site development issues like
  - availability and cost of land,
  - geotechnical conditions,
  - site development,
  - land use,
  - architectural constraints,
  - environmental/permitting requirements,
  - other community concerns (social perceptions and acceptance from communities),
  - availability, cost, and source/type of power and power substation footprint,

- interference with existing infrastructure (intake feed pipe; brine discharge; product line),
- stormwater management,
- tsunami considerations, and
- access.

Development of a desalination project basically involves identifying a water source that can be developed to cost-effectively treat and deliver water to a location where there is a demand or a need for the water. The source water quality typically dictates the level of pretreatment required ahead of the desalination process, while the finished/treated water quality determines the extent of the desalination required (i.e., the number of passes and membrane configuration in an RO system). These factors can have significant impacts on the design and capital cost of the project. Project location and site selection can also greatly influence project costs. Site selection is often a balancing act between finding a location that results in the lowest overall cost for site development, source water intake options, pre- and posttreatment, product water storage and distribution, residuals disposal, and environmental/permitting requirements.

### 5.2.2 O&M Costs

A plant's O&M costs include the costs of actually operating and maintaining the desalination facility and producing desalinated water. These costs typically include power, labor, chemicals, equipment maintenance, media replacement (multimedia filter pretreatment), membrane and cartridge filter replacement, concentrate disposal, repairs, and other replacement parts.

O&M costs may be further broken down into two categories: fixed and variable costs. Fixed O&M costs are those costs that are independent of the amount of water treated and may range from 10–40 percent of the total O&M costs (excluding debt service, which is sometimes considered a fixed O&M cost). Labor costs (including wages/salaries and benefits), repairs, and replacement parts are typically fixed costs.

Variable O&M costs are dependent on and vary with the amount of water that is actually treated at a given time. Power and chemicals are examples of variable costs. The energy costs for desalination plants can vary greatly depending on the salinity of the source water and type of plant, the unit power cost (electricity, oil, or other fuel), the use of energy recovery devices, and other factors. As reported by Bhojwani et al. (2019), energy cost accounts for about 41 percent of the overall O&M costs in SWRO desalination facilities. As a comparison, energy costs represent 66 percent of the overall O&M costs in desalination facilities that use thermal desalination processes. This higher energy demand in thermal desalination is a major driver for the increasing use of SWRO globally.

## 5.3 ESTIMATING CAPITAL COSTS

Capital costs can be broadly divided into direct and indirect capital costs. Direct capital costs are associated with construction of the treatment facilities for production of finished/treated water. Indirect capital costs are other costs that are necessary for the implementation of the project, including engineering, administrative, permitting, land acquisition, and financing costs. This section discusses the key factors that affect capital costs.

### 5.3.2 Direct Capital Cost

Direct capital cost is incurred when constructing the treatment facility. Such costs cover all project components—including civil, mechanical, structural, electrical, instrumentation and control, and architectural work—designed for full functioning of the treatment facility. Below are a number of key factors that affect the design of the treatment facility and other supporting infrastructure and hence the overall direct capital cost.

### 5.3.3 Source Water Quality

Source water quality dictates the pretreatment strategy, as well as the percentage of recovery achievable from the RO process. The ultimate goal for pretreatment is to remove contaminants in source water that could cause rapid membrane fouling and physical damage to equipment (e.g., pumps) or membranes; while the goal of the RO process is to meet specific product water quality goals, particularly dissolved ions that are difficult to remove like boron and bromide. There are three primary types of water quality parameters that are important for determining the pretreatment technology, RO system design and recovery, and longevity and O&M of the RO membranes and therefore affect the overall cost of a desalination project:

- physical parameters: pH, temperature, turbidity, alkalinity, hardness, total dissolved solids (TDS);
- chemical parameters: organic (e.g., biochemical oxygen demand, total organic carbon); inorganic (e.g., iron, boron); volatile organic compounds; synthetic organic compounds; oil; and grease; and
- biological parameters: e.g., coliforms, mussels, algae, jellyfish, and colloidal transparent exopolymer particles.

**Suspended solids and colloids.** Suspended solids and colloids are common contaminants in seawater and are typically removed in the first step of treatment process through chemical–physical processes. Suspended solids and colloids affect treatment facilities more prominently when surface intakes are used but less of an issue for facilities that use subsurface intakes (SSIs). Hence, the type of intakes used plays an important role in determining pretreatment strategies for the treatment plant. The impacts of intake types on pretreatment strategy are discussed later in this chapter.

**Dissolved contaminants.** The TDS level in raw water also affects the energy requirement, RO recovery, and therefore, the overall life-cycle costs of desalination. Additionally, while RO is designed to remove dissolved contaminants, presence of high levels of dissolved organic contaminants can be detrimental to RO membranes as they cause rapid membrane fouling. Fouling not only reduces the run time and water production but also reduces the lifespan of RO membranes. Natural events like algal blooms can increase the level of hydrophilic dissolved organic carbon in water (like proteins and polysaccharides) and can cause rapid membrane fouling if pretreatment is not designed to handle such water quality changes. A typical algal bloom event is characterized by the presence of large amounts of microalgae that alter ambient quality as well as producing toxins and soluble algal products (Seubert and Caron 2018). While RO membranes are effective in removing the toxins produced during an algal bloom event (Seubert et al. 2012) and the product water remains protective of public health, soluble algal products do cause rapid membrane biofouling (Villacorte et al. 2015). Therefore, appropriate pretreatment strategies should be employed to minimize and remove this material from the RO feed water. Some pretreatment options incur more capital cost than others; however, they may also decrease the long-term O&M costs and result in a lower life-cycle cost.

**Oil and grease.** Oil and grease are a concern for RO plants that rely on surface intakes located in shallow waters, particularly along major shipping channels. Oil and grease can foul the membranes rapidly and even cause permanent damage. Pretreatment alternatives like DAF have been incorporated to remove such contaminants to prevent premature replacement of membranes. However, in practice, these systems have rarely been used continuously.

**Temperature.** Source water temperature affects both operating pressures and product water quality. Elevated source water temperature decreases the water viscosity, resulting in lower pressures to push water through the membranes; however, it also increases the osmotic pressure, requiring higher feed pressures. The net impact of temperature on feed pressures will vary, depending on the specific salinity and temperature of the water. Increasing temperatures will increase salt passage, regardless of the salinity, resulting in elevated levels of chloride, boron, and TDS in the permeate. In some cases, the higher salt passage may result in the need for a partial second pass or an increased capacity for the second pass to meet the water quality targets when treating warmer water. Therefore, to achieve permeate TDS targets most efficiently and economically, the operating capacity of the second pass may be varied seasonally.

### 5.3.4 Treatment Capacity and Production Reliability

Treatment capacity and water production reliability determine the size and number of treatment units, pumps, and storage areas for the treatment facility. The costs of larger plants are higher, but they have better economy of scale and lower unit capacity costs than smaller plants. Water production reliability goals determine the level of redundancy required of a treatment plant. The need for high production reliability means that additional treatment trains would need to be installed to provide redundancy to ensure production could be maintained when treatment trains are taken offline for backwash or maintenance purposes or due to unplanned failures. Capital cost for the treatment facilities with high production reliability is higher than those that require lower production reliability. The decision for the level of redundancy needed is determined by a systems engineering analysis that considers the availability of backup supply and costs, product water pricing, storage capacity, the likelihood and consequence of equipment or treatment train failure, and other performance requirements.

### 5.3.5 Raw Seawater Intakes

Seawater intakes can be broadly divided into surface and SSIs. The type of intake can have a significant impact on the cost of desalination facilities, as some types are considerably more costly than others. From a capital cost basis, more costly intakes will often decrease the level of pretreatment required and the long-term operating costs of the facility.

There are many factors that project developers consider before selecting an intake technology for their projects. One of the drivers may be associated with public acceptance and preference or a mandated regulatory requirement. For example, California included specific provisions concerning seawater intake methods in their 2015 Ocean Plan Desalination Amendment. This provision calls for the use of SSIs as the preferred intake option unless project developers are able to demonstrate the infeasibility of adopting SSIs. Screened surface intakes with slot opening of 1 mm and through-screen velocity of no more than 0.5 ft/s (0.15 m/s) could be used as an alternative if SSIs have been sufficiently demonstrated to be infeasible. In California, project developers that opt for the screened surface intake alternative would be expected to invest additional cost to demonstrate such a need and additional time for obtaining such an approval.

Other key intake factors that affect the cost of SWRO desalination projects include the following.

**Surface intakes.** Surface intakes are commonly used for desalination plants, as they are for many other water utility, commercial, and industrial applications. There are three main types of surface intakes: onshore intakes, offshore intakes, and intakes colocated with power generating stations or other industrial facilities.

Onshore surface intakes are typically located along the banks of the source water, while offshore intakes are located away from the shore. Onshore intakes are considerably less costly, yet they are less commonly used for RO-based desalination plants because the near-shore bathymetry may not be favorable; often the water withdrawn from the surf zones has inferior water quality with higher contaminants like turbidity, algae, hydrocarbon, silt, and organics that require higher levels of pretreatment and higher operating costs.

Offshore surface intakes are located as far as several thousand feet from the shore and at least 10 ft (3 m) below the low-tide water level, remaining below the surf zone. Deeper intakes (greater than 65 ft [20 m]) have been proposed to minimize impacts on marine life (USEPA 2014) and improve the quality of water being drawn into the plant; however, depths greater than this are uncommon due to the challenges of conducting maintenance in deep ocean. Water quality tends to be more stable at depths greater than 10–15 ft (3–4.5 m) with less algae, fewer particulates, and fewer surface impacts. At the same time, salinity increases with depth by as much as 1100 mg/L at a depth of 300 ft (91 m) (Dehwah et al. 2015).

In California where once-through cooling for power generation is being phased out, desalination plant developers have considered taking advantage of the abandoned tunnels and converting them into seawater intake and brine discharge tunnels. These efforts have been made more challenging by a 2019 Ocean Plan amendment prohibiting open ocean intakes unless SSIs have been determined to be infeasible. This issue is discussed in greater detail in Chapter 4 of this manual. One major advantage of reusing existing tunnels is that it could minimize offshore construction work needs and seafloor disturbance. The capital cost for construction of screened intakes using this approach is estimated to be US$5–10 million for a facility with 20-mgd (76-ML/d) treatment capacity (Geosyntec 2017).

### 5.3.6 Construction Costs of Surface Intakes

Surface intake construction costs vary depending on site conditions, distance between inlets and the treatment facilities, and complexity of the construction work. It has been reported that the cost range could be 5–30 percent of the total construction costs of the project, with the colocated intakes usually having the lowest construction cost (Voutchkov 2018). Onshore intakes typically have lower construction costs than offshore intakes due to the absence of the offshore construction work component. Based on Poseidon Water's original estimates for the 50-mgd (189-ML/d) SWRO desalination plant in Carlsbad, Calif., in 2007 (Pankratz 2008), the estimated cost for a new intake would be $150 million to withdraw 100 mgd (378 ML/d) of seawater. To retrofit an abandoned conventional surface intake used by a power-generating station with wedge wire screens for a 20-mgd (75-ML/d) SWRO treatment facility, the estimated construction cost was reported to be $6.5 million (in 2017 US$) by West Basin Municipal Water District (Geosyntec 2017). Table 5-1 summarizes some generalized estimated construction cost ranges for surface intakes.

**SSIs.** SSIs involve withdrawing seawater or seawater-influenced groundwater from beneath the ground surface or seabed. SSIs can be broadly divided into types: wells and infiltration galleries. The former involves drilling of wells into geological formation, while

**Table 5-1 Estimated construction cost range of surface intakes (US$)**

| Onshore | Offshore | Colocate | References |
|---|---|---|---|
| • $83,000–92,000 per influent flow in million gal per day (mgd), ±30% | Concrete tunnels<br>• $180–370 per mgd of intake flow per foot, ±30%<br>High-density polyethylene tunnels<br>• $75–130 per mgd of intake flow per foot, ±30% | Around 5% of the treatment facility costs | Voutchkov 2018; Frenkel 2012 |

**Table 5-2 Cost estimates for constructing subsurface intakes (SSIs)**

| SSI Technology | Estimated construction costs of SSIs developed for West Basin Municipal Water District's proposed 20-mgd facility (2019 US$)* | Estimated construction costs of SSIs developed for Carlsbad's 50-mgd facility (2007 US$)† |
|---|---|---|
| Horizontal directionally drilled wells | $80–$120 million, assuming 1–2 mgd per well, with an intake capacity of 18 mgd | Not reported |
| Horizontal wells installed via offshore trenching | $372 million: 14 20-in-diameter microporous horizontal wells; 15 ft–deep trench; 1,850 ft offshore with an intake capacity of 40 mgd | Not reported |
| Seabed infiltration galleries | $278 million for construction costs ($334 million including professional services) | $638 million |
| Slant wells | — | $415 million |
| Ranney collect radial wells | — | $438 million |
| Vertical beach wells | — | $650 million |

* Geosyntec Consultants 2019
† Pankratz 2008

the latter involves installation of collection drains in loose sediment in seafloors or beaches. The two main advantages of SSIs are that they could reduce or even eliminate pretreatment requirements for particulate matter (with the silt density index often well below 5) and avoid entrainment and impingement of marine life (Williams 2018). However, the feasibility of the use of SSIs is determined by site-specific conditions, including hydrogeologic, oceanographic, coastal sediment transportation, and environmental conditions.

### 5.3.7 Construction Cost of SSIs

As mentioned above, the feasibility of using SSIs depends highly on site-specific conditions. Table 5-2 summarizes the construction cost estimates developed for two proposed SWRO desalination facilities in California for installing SSIs.

## 5.4 FINISHED/TREATED WATER QUALITY GOALS

Finished/treated water quality goals are typically dictated by regulations or application-specific contractual requirements. As with any water treatment process, more stringent product water quality goals increase the cost of desalinated water. For example, production of high-pressure boiler feed or green hydrogen production water at an SWRO desalination facility typically involves a two-pass system plus posttreatment, which decreases the overall system recovery, increases capital cost, and requires higher operating costs due to increased power, chemical, and membrane replacement costs. In the case of drinking water, the minimum product water quality goals must meet the applicable drinking water regulations. However, in some cases, utilities may specify a product quality requirement for a specific parameter that is more stringent than the regulatory requirement for drinking water because they may want to more closely match the quality of water from other sources or because of other concerns such as disinfection byproduct formation potential, sodium-restricted diets, or plant irrigation considerations. For these reasons, utilities may specify more stringent limits for TDS, chloride, sodium, hardness, boron, or bromide than that specifically required by drinking water regulations. Meeting more stringent water quality goals may require the use of higher rejection membranes or the use of a partial or complete second-pass SWRO system, which will result in increases in capital and operating costs. Utilities developing desalination projects should examine the costs and benefits of specifying more stringent water quality objectives than required by the regulatory authorities. (For more discussion on water quality goals in desalination, see Chapter 2.)

### 5.4.1 Desalination Plant Provisions for Future Expansion Phase

Phasing strategies can have a significant impact on the capital cost, as plants may be designed with sufficient building space allocation for future RO skids, pretreatment units, or electrical equipment. Similarly, oversized intakes, brine outfalls, or product water conveyance facilities will increase the initial phase of a project. All of these considerations need to be modeled in a cost estimate during the planning phase to determine the best phasing strategy for the project.

### 5.4.2 Finished Water Storage and Conveyance

Along with the seawater intake, storage and conveyance of finished/treated water to customers can result in a significant project cost. The cost of these facilities depends on the onsite storage volume as well as the length, size, and materials of construction of pipelines required to deliver product water to customers. The soil conditions, land use, and population density of the terrain that the transmission main will cross also greatly influence costs. While seawater desalination facilities can often be placed in close proximity to population centers, various factors (e.g., environmental considerations, land availability, supply to inland population centers, and tie-ins to existing infrastructure) have resulted in significant conveyance pipelines for many desalination projects.

In the South Florida Water Management District (CDM 2007) study of nanofiltration and brackish water RO projects, the cost of one half-day of product storage at the rated plant production capacity and conveyance, including high service pumping to the plant site boundary, was only 5–10 percent of the total project cost. In contrast, the cost of the storage and transmission system for a seawater desalination project in southern California in which the product water was to be pumped 20 mi (32 km) into the mountains to tie into the regional water supply was approximately equal to the cost of the desalination process, including pretreatment and posttreatment (Cooley et al. 2006).

These examples illustrate that the cost of product storage and conveyance can vary considerably from project to project. For projects with storage capacity of one half-day of production capacity or less and a short transmission distance, the cost can be as low as 5 percent of total project cost. However, for large storage volumes and long conveyance distances, the cost can approach or exceed the cost of treatment.

### 5.4.3 Concentrate/Brine Disposal

An RO plant produces two streams: a permeate (a water stream that has passed through the membrane) resulting in a higher purity stream from which the majority of dissolved salts have been removed and a concentrate stream that includes the remaining dissolved salts from the original feed stream that remain in the water stream and did not pass through the membrane. For seawater desalination facilities, the TDS concentration of the concentrate is usually 1.5–2 times that of seawater.

Seawater desalination plants almost always return the concentrate back to the ocean or sea in such a way as to facilitate rapid dispersion and assimilation. Concentrate disposal costs are highly variable and depend on the environmental sensitivity in the area of the discharge site and the volume and water quality of the concentrate. (For a discussion on the environmental considerations associated with concentrate disposal, see Chapter 4.)

Simple disposal methods range from a pipe discharging into the surf zone to a complex offshore system with multiple port diffusers. Use of multiple port diffusers allows for increased brine dispersion to reduce potential environmental impacts. The cost of a concentrate discharge system can range from 5 percent of a project's capital cost for a simple channel/pipe arrangement that discharges at the shoreline or surf zone to more than 40 percent of the project's capital cost for a disposal system that discharges far offshore and includes a complex diffuser system. In some cases, existing outfalls that are shared with wastewater treatment facilities or power-generation facilities can be used to provide blending before discharge and reduce the cost of outfall construction.

Because of their design and construction similarities, the cost of a concentrate discharge system is often combined with the intake system cost.

### 5.4.4 Site Development

Some site development issues are straightforward. For example, it can readily be determined whether the parcel of land on which the desalination plant is to be built is of adequate size. The availability and cost of property can be somewhat easily ascertained. Whether constructing a desalination plant is compatible with surrounding land uses is also relatively easy to determine. However, it is more difficult to determine other aspects of construction costs associated with a particular property.

Consider, for instance, geotechnical suitability. One proposed seawater desalination plant was to have been sited adjacent to San Francisco Bay in an industrial area. The property already belonged to the project's owner, was undeveloped, and was large enough for the purpose. However, soil conditions indicated that a pile foundation would have been required and would have added 10 percent to the overall construction cost of the project.

Architectural constraints can also impose cost increases as compared to an industrial building design. Seawater desalination plants by their very nature are located at or very near the seashore. The need for desalinated seawater indicates that the surrounding area is probably developed, perhaps very densely, so locating an industrial type of facility in such an area may be unacceptable. A more elaborate architectural design may be required at some additional cost to get the project permitted.

Power supply can also be a significant cost consideration. Typically, seawater desalination facilities exert large demands on the local power grid. Recently constructed seawater desalination facilities in Australia have employed renewable energy for their primary power supply. The 80-MW Emu Downs Wind Farm in Western Australia, for instance, was built in 2006 for a cost of A$180 million with 66 percent of the power purchased for the 38-mgd (144-ML/d) Perth desalination plant. The water treatment plant itself was built for a cost of A$266 million, suggesting that the power supply for the treatment plant was 45 percent of the cost of the plant. Such power supply costs are generally not included in the reported capital costs of new desalination facilities but are rather incorporated into the O&M costs for purchased power.

### 5.4.5 Indirect Capital Costs

Indirect costs typically include engineering, permitting, legal, financial, and administrative services. The typical indirect costs account for 15–50 percent of the capital costs (Voutchkov 2019). For design–bid–build projects, these costs typically fall to the owner, and bid prices for these projects do not typically include indirect costs; however, utilities should carefully consider these costs in developing a budget for a desalination project. For design–build–operate–finance projects, the majority of these costs will be the responsibility of the project developer and would be included in the cost of water for the project. However, the water purchaser may also incur costs for engineering, legal, permitting, and administrative costs that are not reflected in the cost of water. Each project is unique, and therefore, the magnitude of the indirect costs as a percentage of the construction cost can vary over a very wide range, depending on the size and complexity of the project.

### 5.4.6 Contingency Allowance

Project contingency is generally a function of the level of engineering design that has been completed and the level of detail of the cost estimate. In the initial planning stages of a desalination project, there will be a number of unknowns, things that have not yet been identified. Even for knowns, such as the desalination plant itself, unexpected costs will most likely arise as the project becomes more defined. Therefore, it is necessary to include a contingency allowance in the capital cost budget for the project. At the beginning of a project, the contingency allowance is relatively large. As planning and design proceed, the contingency allowance may be reduced. However, it may be that some of the contingency allowance originally included in the preliminary project budget has been reallocated to construction costs or indirect expenses as these costs were further defined.

The contingency allowance is typically expressed as a percentage of the estimated construction cost, and in the early stages of planning, the contingency allowance may be 25 percent or more of the estimated construction cost. After the final plans and specifications have been prepared, the contingency allowance may be on the order of 10 percent. Even after construction bids are received for the project, a contingency allowance is necessary to cover the unexpected costs that arise on essentially all construction projects.

### 5.4.7 Funding and Financing

Construction of an SWRO desalination facility involves significant capital cost. Although funding the entire project using cash only is possible for certain owners, many seek other funding sources and use cash to supplement the project. Other funding sources that are available for SWRO desalination projects include grants, loans, and/or bond issuance.

Grant funding provided at the federal, state, and local levels is available for SWRO desalination projects that provide drinking water on a competitive basis to supplement the planning and construction costs. For example, the US Bureau of Reclamation provides grant funding for planning and construction of SWRO desalination projects in the western United States and Hawaii through its Title XVI Water Reclamation and Reuse Program. The grant could provide up to 25 percent or US$20–25 million of the planning and construction costs, whichever is lower, to municipal agencies for developing SWRO desalination projects as drinking water supply. State revolving funds also provide grant funding on a competitive basis for SWRO desalination projects. States may allocate some of their state budget to provide grant funding to SWRO project. For example, the California Department of Water Resources has provided multiple rounds of grant funding raised through multiple statewide bond propositions over the years to provide funding for planning and construction of SWRO desalination projects in California.

In terms of financing, there are various types of financing in the form of loans that are available for SWRO projects. Low-interest loans are offered at the federal and state levels. For example, the Water Infrastructure Finance and Innovation Act (WIFIA) loan program administered by US Environmental Protection Agency (USEPA) provides low-interest loans to projects that provide water to public systems. The Drinking Water State Revolving Fund is also a good avenue to obtain ultralow-interest loans for projects that provide drinking water to public systems. Public owners, such as drinking water utilities or municipalities, can also raise capital for project implementation through bond issuance. The interest rates would vary depending on the credit rating of borrowing utilities or municipalities.

Private financing can also be made available through selection of certain alternative project delivery methods, such as design–build–finance–operate (DBFO), design–build–finance–operate–maintain (DBFOM), and other public–private partnerships (PPPs or P3s). The interest rates for DBFOs and DBFOMs are typically specified as part of the contract arrangement. As for other P3 projects, repayment of any interest payment is typically built into the water purchase price as specified in the water purchase agreement.

**Cost of financing calculation example.** In most cases, desalination projects are financed using borrowed money, and interest payments, which can significantly increase the total cost of the project. The following example illustrates the potential significance of the cost of borrowing on the total cost of a desalination project. Begin by assuming the following:

- A 20-mgd (76-ML/d) SWRO project
- Capital cost = $120,000,000
- The desalination plant produces 95 percent of its design annual yield (20 mgd × 365 days/year × 95% = 6,935 million gal/year [26 million $m^3$/year] at rated capacity).
- Financing is obtained for the project with repayment terms for a 25-year loan with an interest rate of 8 percent.

Assuming compounding is done annually, the annual payment would be $11,241,500, and the impact of the financing cost on the cost of water would be $1.62/kgal ($0.43/$m^3$). The cost of financing would be approximately $161,000,000 over the term of the loan.

The previous example is a simple explanation of the financing impact; however, total financing costs vary among projects depending on the complexity, size, location, and financing strategies, including whether the project has government bonds, grants, or direct financing or whether the project is financed entirely through private capital.

Project financing costs include reserves to cover

- interest during construction phase, representing 1.5–4.5 percent of capital cost;
- debt service to protect lenders from a situation where project revenue is insufficient to repay a loan, representing 2–8.5 percent of capital expenses; and
- other financing costs, including insurance reserve; other lender requirements; administrative and legal costs to arrange project loans and equity (applicable for private project finance); working capital required during project construction, startup, and commissioning, as well as during normal plant operations; performance bonds; and taxes, all representing around 0.5–4 percent of capital cost.

Financing costs might represent 4–17 percent of capital costs, depending on the factors mentioned earlier.

### 5.4.8 Construction Costs

Figure 5-1 presents a trendline for unit construction costs based on 62 SWRO desalination plants for municipal drinking water purposes that came on line between 1996 and 2020 worldwide, as reported in the Global Water Intelligence database (GWI 2021). The figure presents a best-fit trendline and adjusted trendlines 30 percent above and below the best fit. Costs have been adjusted to 2020 US$ for each facility, assuming a 3-percent average inflation rate before 2020. These costs do not include indirect (engineering, legal, financial, etc.) costs or contingencies. As reported in Voutchkov (2019), the construction costs for SWRO desalination plants typically account for 50–85 percent of the total capital costs. The figure reflects a historic benchmark, providing a general understanding of desalination market trends; however, it is important to emphasize that desalination costs are highly site specific. High variability of cost is expected, even for similar size capacities, due to the cost

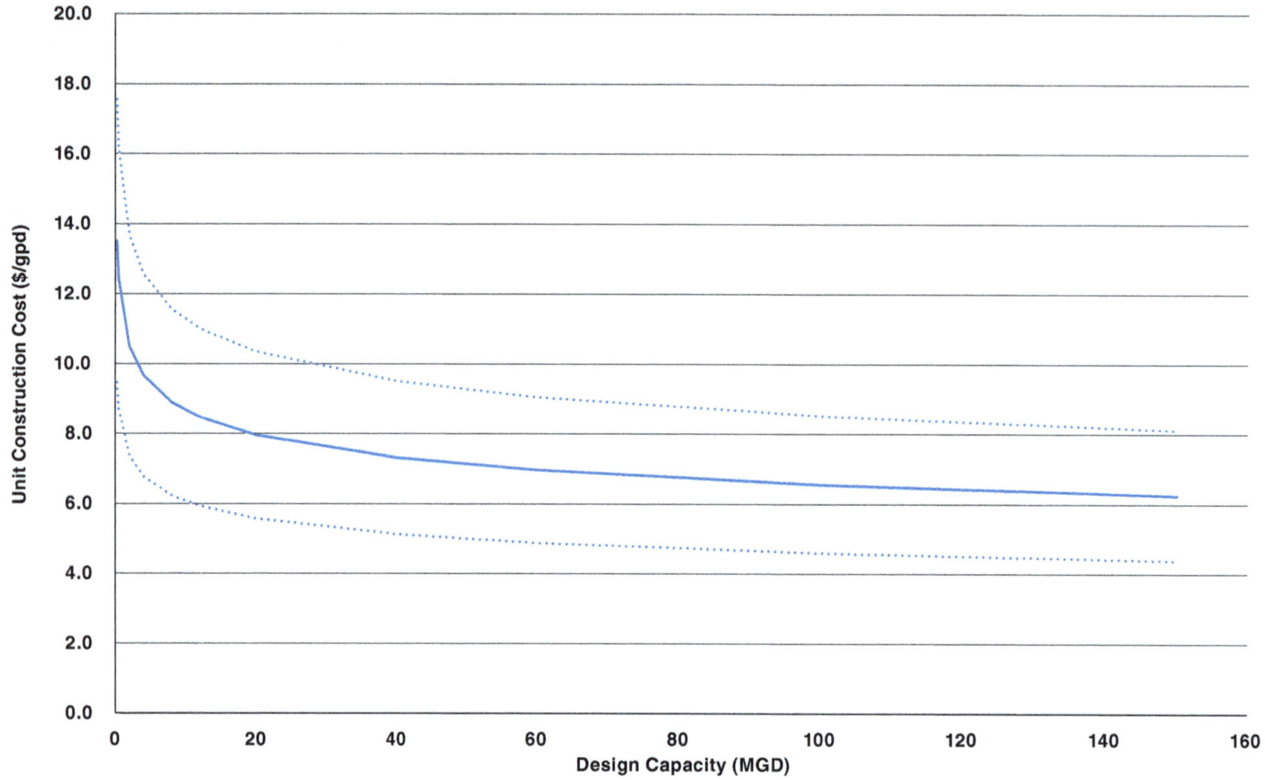


*Adapted from* GWI 2021

**Figure 5-1 Unit construction cost of seawater reverse osmosis (SWRO) desalination facilities worldwide 1996–2020 (2020 US$)**

factors mentioned previously in this chapter, and only 65 percent of the plants in the cost database fall within 30 percent of the best-fit line on Figure 5-1.

## 5.5 PRICE MARKET DISRUPTION AND RECENT COST TRENDS

Recent desalination projects implemented after the global pandemic have experienced a market disruption in materials, equipment, membranes, labor, and logistics that has affected the entire supply chain with delivery time delays, changes to sourcing locations, and overall project costs. The combined effect of both the pandemic and subsequent regional conflicts has been a large price increase in both the capital and operating costs of the projects, which in some cases has increased 30–80 percent compared with cost trends before 2020.

The methodology that has been used by project planners to better understand the recent price increases is discussed in this section. The methodology considers two main inputs: price indices for the main cost items of the desalination plant and price and cost estimates from recent projects contracted after 2020. It may be expected that prices will stabilize to a trend similar to pre-2020 annual changes; however, it is not expected that prices will return to the levels they were before 2020.

### 5.5.1 Price Indices

The main desalination components include mechanical equipment, piping, electrical equipment, and civil works, which can be associated with a price index to characterize

**Table 5-3 Examples of cost indices**

| Cost component | Item | Index used | Variation 2019–2022,* % |
|---|---|---|---|
| Mechanical equipment | Imported | US PPI Electrical Machinery and Equipment | +8 |
| | | US PPI Iron and Steel | +67 |
| Piping | Metallic pipes | US PPI Pressure Pipe and Fittings, Ductile Iron | +26 |
| | | US PPI Steel Pipe and Tube | +79 |
| | | US PPI Steel Pipe and Tube, Stainless Steel | +51 |
| | | US PPI Iron and Steel Pipes and Tubes, From Purchased Iron and Steel | +85 |
| Electrical equipment's | OOEE | US PPI Electrical Equipment Manufacturing | +22 |
| | | US PPI Nonferrous Wire and Cables | +36 |
| | IICC | US PPI Industrial Process Control Instruments | +8 |

*Source:* US Bureau of Labor Statistics
IICC — instrumentation, integration, and control components ; PPI — Producer Price Index; OOEE — owner-or-engineer estimated
* Variation between 2019 and 2022 estimated based on annual variation of prices reported in the PPI, US Bureau of Labor Statistics

price variability over a selected period of time. Each of these cost components can be weighted to the total cost of a project and adjusted separately based on the associated price index change.

Table 5-3 presents an example of how this index can be applied to analyze the price variation of each item in a selected period of time, combining each by their particular weight into the overall cost estimate.

As can be seen, the price increase was significant in almost all the items for years 2019–2022 but varied considerably by the type of component. This analysis demonstrates that using a single inflation factor for the entire facility may not be an adequate approach to estimating costs, given the recent variability in pricing.

For civil works, the *Engineering News-Record* has also published statistics for construction costs, materials, and building indices that can be applied.

### 5.5.2 Procurement and Project Delivery Time Risks

Recent increases and uncertainty in delivery times for equipment and materials create an additional variable in estimating project costs. Project delays can increase costs through schedule inefficiencies or through changes to sourcing of equipment that is found to have unacceptable delays in delivery time.

A method that has been used to reduce this risk is securing critical items with early procurement while identifying the long-lead items that are most likely to impact project schedules. Early procurement of these items minimizes schedule impacts from equipment delays and can reduce the risk of exposure to unforeseen price increases. Contract terms can sometimes limit early procurement; however, this requires a contractor to anticipate future price increases and account for these in the pricing of the project. Doing so can increase the overall cost to the owner. Flexibility in contract requirements can therefore have a considerable impact on facility costs when considerable uncertainty remains in the equipment pricing or delivery times.

### 5.5.3 Recent Desalination Plants Cost Estimates in United States

Two recent desalination project cost estimates in California and Texas are presented in Table 5-4, based on public information released by local water districts. As shown in the

**Table 5-4 Construction cost estimates**

| Characteristics | Estimated Construction Costs of 5-mgd Facility (2023 US$), Expandable to 15 mgd | Estimated Construction Costs of 30-mgd Facility (2024 US$) |
|---|---|---|
| Project name | Doheny Desalination Plant, South Coast Water District (California) | Corpus Christi Desalination plant, Corpus Christi Water (Texas) |
| Intake technology | Slant well | Open intake ship channel |
| Outfall | Dilution with wastewater treatment plant discharge | Discharge ship channel using diffusers |
| Process | Multimedia filter, reverse osmosis (RO), posttreatment, storage, sludge treatment | Dissolved air flotation, ultrafiltration, RO, posttreatment, storage, sludge treatment |
| Construction Cost | $138 million (escalated to 2025) | $541.5 million |

*Adapted from* Clean Energy Capital 2023, Molly 2024

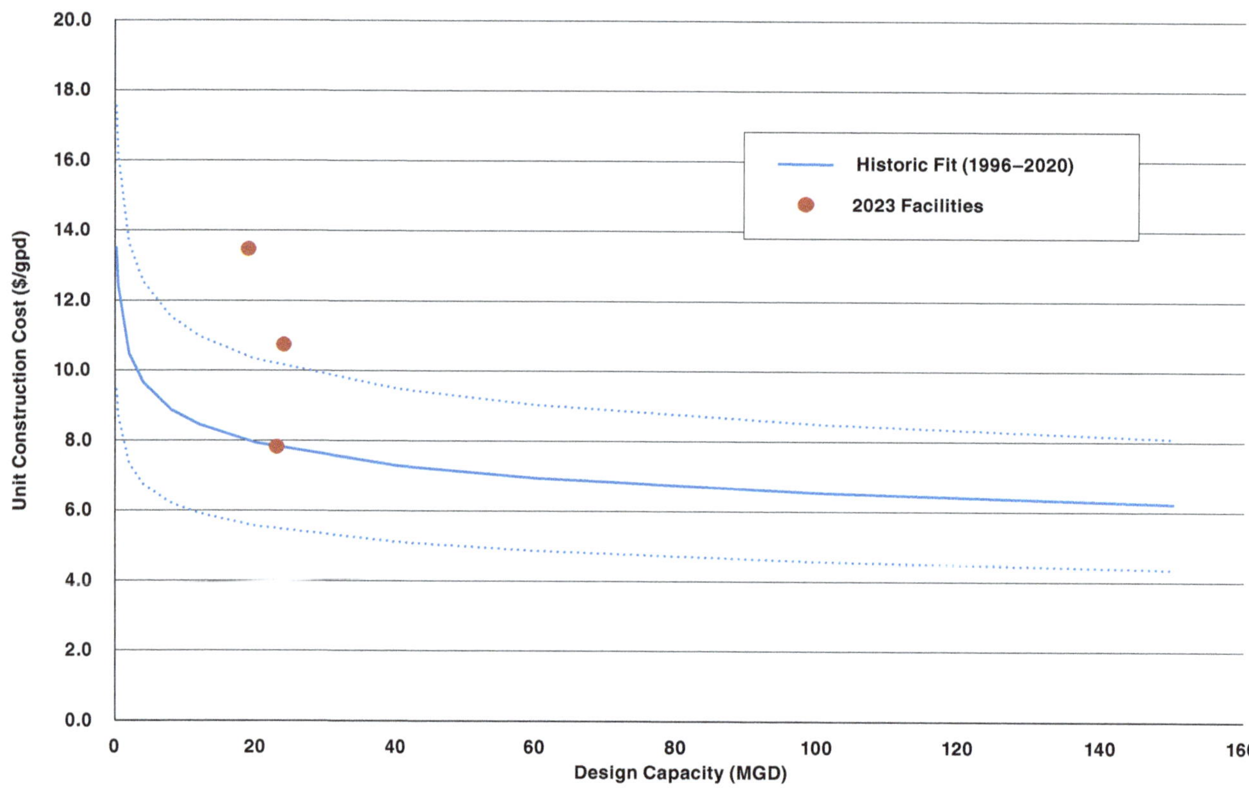


**Figure 5-2 Unit construction cost of seawater reverse osmosis (SWRO) desalination facilities worldwide from 1996–2020 (2020 US$) with recent 2023 facilities**

table, the characteristics of the projects are different due to flow production capacity, location, treatment selection, and current stage of development; however, what these two projects have in common is that the execution delivery method chosen by the utility companies is progressive design-build (see further references in Water Collaborative Delivery Association 2023). It should be noted that the unit costs for the two proposed US facilities listed in Table 5-4 are 4–5 times higher than the historic best-fit trendline presented previously in Figure 5-1.

Beyond the recent estimates for these two US facilities, Figure 5-2 incorporates three international 2023 project costs with the historical data from 1996–2020. These more recent numbers suggest that even outside the United States, a significant cost increase has occurred compared with historical trends.

## 5.6 ESTIMATING O&M COSTS

With respect to O&M costs, labor is typically considered a fixed O&M cost. Wage levels may change with time, but labor costs are not directly impacted by the amount of water that is desalinated. Another fixed O&M cost is RO membrane replacement.

Generally, once water has passed through a membrane, the membrane is considered to have begun its useful life. In general, membranes are assumed to age at a constant rate regardless of if they are in service or not. For example, if a membrane plant is operated only 50 percent of the time, it is not generally assumed that the membrane's life would be twice as long as it would be if the plant was operated 100 percent of the time. In reality, membranes that are used less frequently tend to last longer than those in continuous operation, and off-line membranes stored in permeate rather than feed water can last considerably longer.

Some allowance should be included for repairs and replacement parts as a fixed O&M cost for planning purposes; an allowance of 1–2 percent of the plant construction cost per year is typically included to cover this. There may be other fixed O&M costs specific to a given project. Fixed O&M costs are usually evaluated on an annual cost basis.

Variable O&M costs are those costs that directly depend on the amount of water that is desalinated. Variable costs may include power, chemicals, concentrate disposal, and other miscellaneous costs. The total annual O&M cost is the sum of the fixed and variable O&M costs in dollars per year.

### 5.6.1 Fixed O&M Costs

**Labor and labor overhead.** With regard to labor, the cost is dependent on a number of factors, including the following.

*The size (capacity) of the plant.* A small plant may operate with only one or two permanent staff members with unattended operation during some periods of the day. In contrast, larger plants typically require operators in attendance at all times to keep up with routine maintenance, record-keeping, etc.

*The complexity of the plant.* If the treatment process is complicated, it may be necessary to have a larger operating staff to cover all the skills needed for plant operation.

*Regulatory agency requirements.* Local or state regulatory agencies may have minimum staffing, certification, and training of the personnel as requirements for operating the desalination plant.

*Owner's policy.* Some desalination plant owners may desire to have at least one operator at the plant around the clock, while another owner may wish to have as small an operating staff as possible with the plant operating unattended at times.

*Local wages/benefits.* Some areas, particularly in or near densely populated metropolitan areas, may have higher wage/benefit costs than more rural areas.

**Membrane replacement.** The cost of membrane replacement is usually regarded as a fixed cost because once a membrane is placed in operation, it is generally assumed that membrane degradation begins. A membrane replacement fund may be established to accumulate money to replace membranes as needed. The two primary factors considered in establishing a membrane replacement fund are the total cost of the membranes and the expected average life of the membranes.

Determining the initial cost of the membranes depends on several factors, including

- the capacity of the desalination process,
- the number of treatment passes required,
- the recovery expected,
- the membrane design flux, and

- rotation regimen (replace new membranes and rotate old ones) to prolong lifespan.

In addition, to account for future price increases, it is important to index the price of membranes during the O&M period.

Estimating the annual membrane replacement cost is basically a function of the number of membranes installed, the projected membrane replacement cost, and the anticipated useful life of the membranes. The useful life of the membranes primarily depends on the feedwater quality and the effectiveness of pretreatment. For this reason, the life expectancy of second-pass membranes is typically longer than first-pass membranes. Membrane life expectancy has been increasing from a traditional life expectancy of 3 years to the point that life expectancies of 5–7 years are not unusual, especially with MF/UF as pretreatment.

An allowance for repairs and replacement parts beyond membrane replacement should also be included when planning a desalination project. The cost for repairs and replacement parts is typically taken as 1–2 percent of the plant construction cost.

## 5.6.2 Variable O&M Costs

**Power.** All of the power required by the desalination project should be included in O&M when estimating the cost. For instance, the power required to pump the feedwater to the plant from the source(s) of supply and the product water to customers should be accounted for as well as the power required for the desalination process itself. The power required (kW of demand and kW•h consumed) can usually be relatively accurately defined in the preliminary phases of project planning.

It is important to obtain as accurate an estimate of power costs ($/kW•h) as possible because power is usually a substantial portion of the O&M cost. This is especially true for seawater desalination plants because they consume more power than brackish water treatment facilities. Most of the power cost will be a variable cost. However, a small percentage will be a fixed cost because even when the desalination system is not operating or is operating at reduced capacity, there will be some power demand for building loads and site lighting.

**Chemicals.** The types and amounts of chemicals that might be required in a desalination plant can vary significantly. Chemicals are required for the following purposes:

*Filtration.* If particle removal is needed ahead of the desalination process, chemicals may be necessary to increase filter performance. Such chemicals may include coagulants, filter aids such as polymers, acid for pH adjustment, and disinfection chemicals to control biogrowth. In case of UF pretreatment, it is common to conduct periodic chemically enhanced backwashes using sodium hypochlorite, sodium hydroxide, or citric acid to maintain membrane performance. Spent cleaning solution must typically be neutralized, further adding to chemical usage.

*Desalination process feedwater.* Historically, it has been common practice to add a scale inhibitor and/or acid ahead of the SWRO membranes to inhibit calcium carbonate scale formation and reduce the risk of scaling/fouling on the membranes. More recently, SWRO systems with TDS of 35,000 mg/L and pH of 8.0 or less have been successfully operated without acid and antiscalant addition in the recovery range of 45–50 percent. With higher efficiency energy recovery systems, designers sometimes opt for slightly lower recoveries in the 45–47 percent range to allow some additional safety margin with respect to scaling potential. Operation at higher recoveries will still likely require the addition of acid and/or scale inhibitor to avoid scale formation.

*Posttreatment.* SWRO plants typically reduce hardness and alkalinity to very low levels, which results in a relatively high corrosion potential of RO permeate. To compensate for this, it is common practice to add hardness and alkalinity to stabilize the RO permeate. Lime (calcium hydroxide) or calcite (calcium carbonate) is typically added to increase the hardness of the finished water. Mineral acid or carbon dioxide is also added to react with the lime or calcite to produce alkalinity. A disinfection residual in the form of chlorine or chloramines is also added after desalination. The selection of the disinfectant type is often based on the disinfection practices used for water from other sources that may be blended with finished water from the desalination facility.

*Membrane cleaning.* RO membranes require periodic cleaning where the frequency of cleaning and the type of chemicals required vary depending on the particular case. Generally, a low-pH (acid) solution is used to clean mineral scales, and high-pH (caustic soda) solution is used to clean biological fouling. In addition, a detergent may be used from time to time. If membrane filtration (MF/UF) is used to filter the desalination process feedwater, more chemicals will be used to clean the pretreatment membranes.

The cost for chemicals depends on the quality and quantity of the water being treated and the product water quality goals. Chemical consumption data from other SWRO plants treating seawater of similar quality can be used to estimate chemical treatment needs and costs for a new desalination facility. However, if such data are not available, pilot testing is typically performed to better quantify chemical requirements and subsequent costs, particularly for very large SWRO systems.

**Residuals disposal.** Desalination plant residuals include the concentrate stream and residual solids that were removed from the raw water before membrane treatment. Residual solids include suspended solids along with coagulation chemicals that were used in the pretreatment process. For any desalination process, it is important to develop a strategy for concentrate volume reduction and disposal and to determine the associated costs and permitting requirements in the planning stages of the project.

For seawater desalination projects, concentrate is usually returned to the ocean. If sources for dilution of the concentrate are available, such as cooling water from a power plant or ocean discharges from a wastewater treatment plant, dilution of the concentrate flow by mixing with these streams can facilitate the permitting process and reduce the cost of permitting. In addition, use of existing outfall structures can significantly reduce the cost of concentrate disposal. Where dilution alternatives are not readily available, installation of diffusers can be used to promote rapid mixing and dispersion of the concentrate. Diffusers can reduce discharge concentrations close to background ocean salt concentrations within relatively short distances from the discharge points. The costs associated with additional pressure drops across the diffusers to promote mixing and the energy cost associated with discharging concentrate farther away from the plant site and/or further offshore to take advantage of prevailing ocean currents to further promote mixing should be considered in the O&M costs. Costs for monitoring water quality in the discharge area to assure compliance with permitting requirements is an important factor to account for in the O&M costs.

The cost of residuals disposal from the pretreatment system can vary widely depending on raw water quality and the options for disposal of residual solids. For installations where suspended solids levels are relatively low and where membrane filtration can be used with low dosing of coagulants or even without coagulants, it may be possible to return residuals to the ocean with minimal treatment costs. On the other hand, if the desalination plant is close to an estuary where high runoff flows and high turbidity can be experienced, the amount of solids may be too great for permittable discharge. If beneficial uses for these residuals cannot be found, disposal of these solids in a landfill may be required. Disposal costs for landfills can vary significantly from location to location, and in some areas, disposal costs may be high and can become a significant O&M cost. There are also treatment costs associated with dewatering (filter press or centrifuge) that should be considered in the development of O&M cost estimates. For these reasons, residuals management should be carefully evaluated in the development of a desalination project.

## 5.7 COST OF WATER

The sum of the annual capital repayment and annual O&M cost divided by the volume of water produced during the year is termed *the cost of water* and is usually expressed as dollars per unit volume: for example, in dollars per 1,000 gal ($/kgal), dollars per cubic meter ($/m$^3$), or dollars per acre-ft ($/acre ft).

Total production costs are dependent on the specific site and application, treatment processes used, plant facilities constructed, power, labor, and chemical costs, treatment plant capacity versus annual water production, and other factors. Table 5-5 presents three class ranges of water production costs of 50 medium and large SWRO desalination facilities worldwide reported by Voutchkov (2019) and adjusted to 2020 dollars. Based on the dataset evaluated by Voutchkov (2019), the cost breakdown reported is as follows:

- direct capital costs: 30–40 percent,
- indirect capital costs: 10–20 percent,
- power cost: 20–35 percent, and
- other O&M costs: 15–30 percent.

The numbers in Table 5-5 come from reported costs from facilities around the world; however, vast differences are seen in the cost of water for facilities in different geographic regions. Facilities in Asia and the Middle East have historically had the lowest costs, with the cost of water often below US$2.0/kgal (as reported by GWI and the Water Desalination Report from Media Analytics). Facilities in Australia have had the highest unit costs, exceeding US$6.0/kgal. The small number of facilities operating in the United States makes it difficult to report a typical cost; however, with the complex regulatory environment in the United States (discussed in Chapter 4), costs similar to or exceeding those in Australia should be anticipated.

### 5.7.1 Cost of Water Estimate Example

Tables 5-6 (estimated capital cost), 5-7 (estimated O&M costs), and 5-8 (estimated water costs) present an example of how a preliminary cost estimate for a desalination project might be calculated. The example is based on

- finished/treated water design capacity = 50 mgd (189 ML/d);
- annual production rate based on
  - 100 percent of rated plant capacity = 18,250 million gal per year (MG/year) or 69 million m$^3$/year and
  - average day demand at 80 percent of rated capacity = 14,600 MG/year (55 million m$^3$/year);
- desalination process recovery of 50 percent; and
- a single-pass design.

**Table 5-5 Water production costs for medium and large seawater reverse (SWRO) osmosis desalination facilities worldwide**

| Class Range | Cost of Water Production (US$/kgal, 2020 US$) |
|---|---|
| Low-end range | $1.97–3.15 |
| Midrange | $3.55–5.91 |
| High-end range | $6.30–11.82 |
| Average | $4.33 |

*Adapted from* Voutchkov 2019

The rule-of-thumb range method presented by Voutchkov (2019) posits two scenarios of costs and also applies here:

- Low range in capital and operational cost, which considers an onshore intake and typical outfall, pretreatment based on multimedia gravity filter, reverse osmosis, and posttreatment with a calcite contactor. For the O&M, this scenario assumes average conditions for labor, maintenance, chemicals, replacements, and average specific energy consumption of 12.1 kW•h/kgal (3.2 kW•h/$m^3$), with an energy price of 0.1 $/kW•h.
- High range in capital and operational cost, which considers offshore open intake based on concrete tunnel, pretreatment based on DAF and a multimedia filter, and posttreatment with a calcite contactor. For the O&M, this scenario assumes a higher range for labor, maintenance, chemicals, replacements, and energy consumption of 13.2 kW•h/kgal (3.5 kW•h/$m^3$), with an energy price of 0.1 $/kW•h.

(All prices were updated in 2023 US$.)

Table 5-6 shows the impact on the capital cost due to technical differences mentioned earlier in this section related to the type of intake and pretreatment selection where the low range cost scenario considers typical conditions for an SWRO plant while the high range cost scenario accounts for a complex intake and difficult pretreatment.

Table 5-7 illustrates how O&M costs for this desalination plant might be presented. Two annual production rates are provided to show the impact on the cost of water by spreading the fixed costs of O&M over a greater volume of water production. As shown in the table, the variable O&M costs were adjusted to reflect the fact that if more finished/treated water is produced, the annual cost for power and chemicals will increase correspondingly. Annual fixed costs do not change. The result is that the unit O&M cost per 1 kgal of water produced at the annual production rate of 18,250 MG/year is lower than the unit O&M cost for producing only 14,600 MG/year from this facility.

**Table 5-6 An example of a seawater reverse osmosis (SWRO) plant capital costs**

| | Low Range | High Range |
|---|---|---|
| Plant Capacity: 50 mgd (189 ML/d) | 2023 | 2023 |
| **Cost item** | | |
| **Direct capital (construction) costs** | | |
| 1. Site preparation, roads, and parking | $3,785,412 | $4,353,224 |
| 2. Intake | $24,050,000 | $120,310,125 |
| 3. Pretreatment | $26,500,000 | $64,369,999 |
| 4. Reverse osmosis system equipment | $88,200,000 | $101,430,000 |
| 5. Posttreatment | $4,830,000 | $5,554,500 |
| 6. Concentrate disposal | $7,350,000 | $36,768,375 |
| 7. Residuals management | $3,785,412 | $4,353,224 |
| 8. Electrical system | $7,570,824 | $8,706,447 |
| 9. Product storage and distribution | $5,678,118 | $6,529,835 |
| 10. Buildings | $11,356,235 | $13,059,671 |
| 11. Startup | $3,785,412 | $5,678,118 |
| **Subtotal direct (construction) costs** | **$186,891,412** | **$371,113,517** |
| Project engineering services | 30,232,434 | 58,960,000 |

*(continued)*

**Table 5-6 An example of a seawater reverse osmosis (SWRO) plant capital costs (*Continued*)**

| | Low Range | High Range |
|---|---|---|
| Plant Capacity: 50 mgd (189 ML/d) | 2023 | 2023 |
| Capacity m$^3$/day | 189,271 | — |
| **Project Development Costs** | | |
| 1. Administration, contracting, and management | 5,678,118 | |
| 2. Environmental permitting | 5,678,118 | |
| 3. Legal services | 2,839,059 | 6,624,471 |
| **Subtotal project development** | **14,195,294** | **30,283,294** |
| Project financing costs | 25,695,000 | 51,147,000 |
| Contingency | 28,550,000 | 56,830,000 |
| Subtotal indirect capital costs | 98,672,729 | 197,220,294 |
| Total capital costs | **285,564,141** | **568,333,812** |
| Unit cost, $/gpd | $5.711 | $11.367 |
| Unit cost, $/m$^3$/d | $1.509 | $3.003 |

*Source:* Ruben Muñoz

**Table 5-7 An example of annual operation and maintenance (O&M) costs for treatment technology: seawater reverse osmosis (SWRO)**

| Item No. | Description | Plant Capacity (mgd) 50 Low range | High range |
|---|---|---|---|
| | **Fixed** | | |
| 1. | O&M labor | $2,072,513 | $3,108,769 |
| 2. | Replacement parts and materials | $3,454,188 | $4,835,864 |
| 3. | Replacement membranes | $1,924,750 | $2,114,021 |
| 4. | Environmental and performance monitoring | $656,296 | $898,089 |
| 5. | Indirect O&M cost | $2,763,351 | $4,145,026 |
| | **Subtotal Fixed** | **$10,871,098** | **$15,101,769** |
| | **Variable Costs** | | |
| 6. | Power cost $ /year | $22,106,805 | $29,015,181 |
| 7. | Chemicals, sludge disposal, variable maintenance, other variables cost, $/year | $3,799,607 | $7,737,382 |
| | **Subtotal Variable $ /year @ rated capacity** | **$25,906,412** | **$36,752,563** |
| | Annual production at rated capacity (million gal/year) | 18,250 | 18,250 |
| | **Annual O&M cost at rated capacity** | **$36,777.510** | **$51,854,332** |
| | Unit cost at rated capacity, $/kgal | $2.02 | $2.84 |
| | Unit cost at rated capacity, $/m$^3$ | $0.53 | $0.75 |

*(continued)*

**Table 5-7 An example of annual operation and maintenance (O&M) costs for treatment technology: seawater reverse osmosis (SWRO) (*Continued*)**

| Item No. | Description | Plant Capacity (mgd) 50 Low range | High range |
|---|---|---|---|
| | Average day capacity/rated capacity | 80% | 80% |
| | Annual production at average day demand (million gal/year) | 14,600 | 14,600 |
| | **Annual O&M cost at average day demand capacity** | **$31,596,227** | **$44,503,819** |
| | Unit cost at rated capacity, $/kgal | $2.16 | $3.05 |
| | Unit cost at rated capacity, $/m$^3$ | $0.57 | $0.81 |

*Source:* Ruben Muñoz

**Table 5-8 An example of total costs water treatment technology: seawater reverse osmosis (SWRO)**

| Item No. | Description | Plant Capacity (mgd) 50 Low range | High range |
|---|---|---|---|
| **Production Costs at Rated Capacity** | | | |
| 1. | Equivalent annual capital cost | $26,751.300 | $53,240,818 |
| 2. | Annual operations and maintenance (O&M) cost—fixed | $10,871.098 | $15,101,769 |
| 3. | Annual O&M cost—variable | $25,906.412 | $36,752,563 |
| Total annual cost: | | $63,528.810 | $105,095,149 |
| | Annual production at rated capacity, (million gal/year) | 18,250 | 18,250 |
| | **Annual production cost at rated capacity ($/kgal)** | **$3.48** | **$5.76** |
| | **Annual production cost at rated capacity ($/m$^3$)** | **$0.92** | **$1.52** |
| | Average day capacity/rated capacity | 80% | 80% |
| 1. | Equivalent annual capital cost | $26,751,300 | $53,240,818 |
| 2. | Annual O&M cost—fixed | $10,871,098 | $15,101,769 |
| 3. | Annual O&M cost—variable | $20,725,130 | $29,402,050 |
| Total annual cost: | | $58,347,527 | $97,744,637 |
| Annual finished water production rate (MG/year)(3): | | 14,600 | 14,600 |
| **Annual production cost at average day demand ($/kgal):** | | **$4.00** | **$6.69** |
| **Annual production cost at average day demand ($/m$^3$):** | | **$1.06** | **$1.77** |

*Source:* Ruben Muñoz

Table 5-8 compares the cost of water for the same annual production rates. The annual capital repayment cost is constant, as are the fixed O&M costs. Only the variable O&M costs change with the change in annual production rate. As the total cost of water figures in the table show, producing more water from the desalination plant reduces the cost of water.

The annual capital cost was calculated using an interest rate of 8 percent and a 25-year amortization period (annual debt payment = 0.0937 × $285,564,000).

## REFERENCES


Bhojwani, S., K. Topolski, R. Mukherjee, D. Sengupta, and M.M. El-Halwagi, 2019. "Technology Review and Data Analysis for Cost Assessment of Water Treatment Systems." *Science of the Total Environment*, 651:Part 2:2749–2761.

CDM. 2007. *Water Supply Cost Estimation Study: South Florida Water Management District. 2007.* www.sfwmd.gov/sites/default/files/documents/water%20supply%20cost%20estimation%20study%202-2007_cdm.pdf (accessed October 20, 2025).

Clean Energy Capital. 2023. *Cost of Water Update and Rate Impact Analysis: South Coast Water District Doheny Desalination Plant.* https://cms9files.revize.com/scoastwaterdist/9a_-_2023-06-16_Doheny_Desal_-_CEC_Board_Slides.pdf (accessed November 20, 2025).

Cooley, H., P. Gleick, and G. Wolff. 2006. *Desalination, with a Grain of Salt: A California Perspective.* Oakland, CA: Pacific Institute.

Dehwah, A., S. Li, S. Al-Mashharawi, F. Mallon, Z. Batang, and T. Missimer. 2015. "Effects of Intake Depth on Raw Seawater Quality in the Red Sea." *Environmental Science and Engineering*. v149. Berlin: Springer Nature: 105–124.

Frenkel, V.S. 2012. *Consideration for the Co-Siting of Desalination Facilities with Municipal and Industrial Facilities – Final Project Report and Decision Tool.* Alexandria, VA: WateReuse Research Foundation.

Geosyntec Consultants. 2017. *Seabed Infiltration Gallery Construction and Life-Cycle Costs for a Proposed 20 MGD Ocean Water Desalination Facility in El Segundo, California.* Draft Environmental Impact Report. West Basin Municipal Water District.

Geosyntec Consultants. 2019. *Supplemental Feasibility Assessment of Subsurface Seawater Intakes Along the Santa Monica Bay Coast for the Proposed Desalination Facility in El Segundo, California*. Final Environmental Impact Report. West Basin Municipal Water District.

Global Water Intelligence (GWI). 2021. Water Data. https://www.gwiwaterdata.com/ (accessed November 20, 2025).

Molly, P. 2024. *Inner Harbor Seawater Desalination Treatment Plant Update. Corpus Christi Water.* https://0673fdba3b9842f385b55dcb4be1141d.usrfiles.com/ugd/0673fd_6b41e481e2d9445db465cd5610eff4f9.pdfP (accessed November 29, 2025).

Pankratz, T. 2008. "Water Desalination Report." 44(33). https://www.desalination.com/.

USEPA (US Environmental Protection Agency). 2014. *Final Regulations to Establish Requirements for Cooling Water Intake Structures at Existing Facilities and Amend Requirements at Phase I Facilities*. Washington: USEPA.

Water Collaborative Delivery Association. 2023. *Progressive Design-Build Procurement Guide.* Denver: WCDA.

Williams, D. 2018. "Design and Construction of Subsurface Intakes." *Sustainable Desalination Handbook: Plant Selection, Design and Implementation*. V. Gnaneswar Gude (Editor). Oxford: Butterworth-Heinemann:227–258.

Voutchkov, N. 2019. *Desalination Project Cost Estimating and Management*. Boca Raton, FL: CRC Press.

Voutchkov, N. 2018. "Design and Construction of Open Intakes." *Sustainable Desalination Handbook: Plant Selection, Design and Implementation*. V. Gnaneswar Gude (Editor). Oxford: Butterworth-Heinemann:201–255.

This page intentionally blank.

Chapter 6

# Construction and Procurement of Desalination Facilities

*Val S. Frenkel, Peter Walkemeyer, Chris Morss, Richard Whiting*

## 6.1 INTRODUCTION

Water supply has become a critical issue worldwide during the last decade. As a result, seawater desalination has rapidly expanded to become one of the major sources of drinking water for many regions and countries. Figure 6-1 provides a breakdown of desalination capacity by continent. Desalination projects often require development to occur within short timeframes using complex technologies that continue to evolve. The variables and risks associated with seawater desalination plant developments can be an enormous challenge. Public and private utilities often consider alternative project procurement strategies for major desalination projects, which is one of the key reasons that desalination projects are considered to be high-risk projects when compared to traditional water treatment facilities. Many seawater desalination projects around the globe are procured using the alternative project delivery approaches described in detail in this chapter. The ultimate selection depends upon the project size, location, schedule, in-house capabilities, experience of the public entity, and financing available to the public entity.

Many methods of project procurement and delivery exist for delivering a successful water project. Consideration of each will inform the owner which path will best serve their needs. This section addresses the fundamentals of the most prevalent methods that have been used and provides a comparison of the advantages and disadvantages to each. While this is a compact listing of methods, many more exist that are variations on the methods described here.

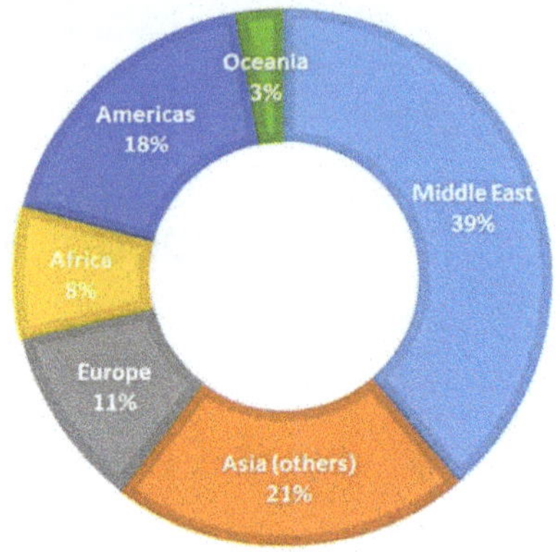


(a)

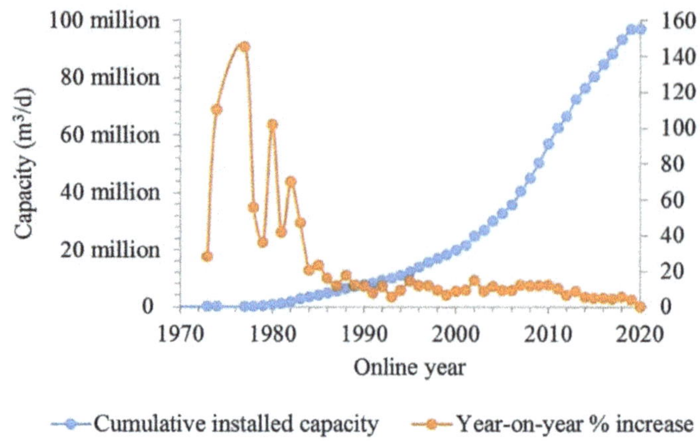


(b)

*Adapted from* Global Water Intelligence 2020

**Figure 6-1 Desalination capacities by continents (a) and percentage by continents out of total capacity 97M m$^3$/d (b) annual increase**

Before selecting a procurement and project delivery method, a public entity (and its advisors) should consider several factors that set the foundation for the project and allow informed decision-making regarding project delivery. Beginning a project with a solid understanding of needs and capabilities before selecting a delivery method will aid in selecting the most appropriate for each project situation. Evaluating these factors should identify the criteria that are most important to the project's success, measured in terms of budget, affordability, timely delivery, and long-term reliability, flexibility, and functionality.

The public or private entity should perform a conceptual design and establish its budget as soon as possible in the process. These are most commonly developed by the entity contracting an outside consultant engineer with experience in seawater desalination project development. During the process, the owner should take into consideration the potential for both budget and schedule changes once a project is underway, which can be costly if not well planned. Consideration of the option for a public–private partnership at this stage is critical as the cost structure and financing approach will be very different using a private sector owner-operator of the plant and will significantly reduce the project capital budget required from the public sector partner during the construction phase, although possibly increase the costs during the operating phase. It also significantly reduces the risks of budget and schedule overruns for public utilities.

An accurate construction schedule is also critical for overall project completion and results. Sometimes schedules are estimated ambitiously because of public or corporate pressure to alleviate water shortages, which can set up a project for overruns—with their associated cost impacts—and unhappy stakeholders. It is much easier to avoid disputes and claims if plans and schedules are agreed to by all and firmly in place before breaking ground.

A thorough risk evaluation should be conducted before a project begins because many construction risks can result in increased costs, delays, and even project failure. Determining who will be liable for problems that arise from design, engineering, or construction that result in damages during and beyond construction may help prevent disputes and possible litigation. Additionally, if the public entity considers a project delivery method where in-house resources will be involved in project administration, it must also consider its responsibility for reducing gaps in design, engineering, and construction services. A public sector utility or a private corporation passing on unquantifiable risks (such as unknown soil conditions and/or unknown intake seawater quality) to a private contractor to fix will inevitably lead to high construction pricing, conditional pricing (with the potential for later claims for extra costs), or disputes.

Finally, the public entity must assess its previous experience in the construction of previous projects of similar scope and size. Ensuring that a project has the right amount of personnel with the appropriate expertise will set the project up for success. While extensive construction expertise is not crucial to a project's success, it will certainly help inform and assure the results by the selected delivery method (see Figure 6-1 for further insight).

## 6.2 CONSTRUCTION AND PROCUREMENT METHODS

A seawater desalination project will be successful if it meets the business and project objectives, provides value for money, and effectively manages risk. The delivery strategy selected for each project needs to support the achievement of these objectives and deal most appropriately with the risks. It also needs to work within the constraints and suit the level of project complexity. Selecting a contract delivery method aligned to the project objectives is essential to the ultimate success of this project. There are many project delivery methods used worldwide, and methodologies may be country specific. The two major methods to procure and deliver projects are the following:

- Conventional procurement, known as design–bid–build (DBB), and
- Alternative procurement, which consists of several options.

### 6.2.1 Conventional Procurement

The conventional procurement methodology (DBB) is the typical process routinely used by public (government-owned) utilities (the project owner) to achieve project delivery. The public utility funds 100 percent of the project development, design, and construction, and adopts the completed asset once the identified and associated acceptance tests have been completed satisfactorily. The key item to note is that all project risk remains with the public utility, including delivery program, budget, and ultimate asset performance.

DBB is a robust, verified process wherein the design phase of the project, from concept to 100 percent design completion, is typically completed by an engineering company appointed by the public utility via a separate qualification and bidding process. The next step is a bidding process to select a construction contractor using the standard request for qualifications (RFQ) and request for proposals (RFP) process with documents produced by the engineer based on the detailed design and specifications developed during the design phase. The contractor therefore must respond to a detailed description of the work. The benefit of this procurement strategy is the public utility maintains maximum control over the project and expects delivery of the project as it was designed and specified. The major drawbacks of a conventional procurement methodology include

- the multiphase sequential staging of the project, which requires a longer schedule to deliver the project when compared to alternative project delivery models;
- the engineer taking limited or no responsibility for cost overruns due to design deficiencies; and
- the requirement for the public utility to have at least an outline knowledge and understanding of every design, construction, and operations aspect of the project, including the latest technological advances, which are often in the domain of specialist process providers and contractors.

### 6.2.2 Alternative Procurement

Alternative project delivery models are different from DBB in that they may typically require different arrangements between the parties when developing a particular project. While conventional procurement consists of one general DBB procurement method, the alternate procurement has several variations. The main alternative delivery methods are the following.

- **Design and construct:** design, develop, and construct (DD&C), design–build (DB), design and construct (D&C), engineer–procure–construct (EPC), or design–construct–maintain (DCM)
- **Management:** construction management, engineering–procurement–construction–management (EPCM), cost plus, or cost reimbursable performance incentive
- **Relationship:** contracting, construction management at risk (CMAR), managing contractor, early contractor involvement (ECI), or progressive design–build (PDB)
- **Financed:**
    - design–build–operate (DBO), build–transfer–operate (BTO), design–build–operate–maintain (DBOM), build–operate–transfer (BOT), build–own–operate (BOO), design–build–finance–operate (DBFO), design–build–finance–maintain (DBFM), design-build-finance-operate-transfer (DBFOT), build–own–operate–transfer (BOOT), finance–design–build–own–operate–transfer (FDBOOT); and
    - Public–private partnership (PPP or P3)
- **Alliance method**

Not many engineers are experienced in alternative project delivery models that include private sector financing. As a result, they tend to overestimate the appeal of the low-percentage equity returns for the private sector while underestimating the cost of private sector debt, the impact of budget and program overruns on the cost of low-interest public debt or on higher-interest private sector debt, and the value to the public entity in the transfer of process and long-term operations risk to the specialized private partner. If a public sector entity is interested in considering such a project delivery model, it should ensure that it is receiving the analysis from qualified advisers, including financial advisors who are well experienced in private sector financing and in operations and maintenance (O&M) costs. The public sector entity may also want to include potential private partners in these early discussions.

### 6.2.3 Detailed Descriptions of Procurement Strategies

More detailed descriptions of procurement strategies are presented below.

**DBB.** DBB is the most traditional method to deliver a project and one that is well accepted by owners, designers, engineers, and contractors. This method aims to secure the best price for a given construction project by limiting the financial risk for contractors bidding/executing the construction phase of the project and also by not limiting the contractors bidding to those with design capability. This project delivery method consists of the owner contracting with separate parties for design/engineering the project and for construction of the project. There are three main sequential phases to the DBB delivery method: the design phase, the bidding phase, and the construction phase.

The owner initially works with the design engineer to design the project. It is crucial to keep to the overall budget for the project that the design engineer is contracted to carry out detailed designs with very little interpretation left to the construction contractor. Seeking the lowest cost design engineer and lowest cost design at this stage is inadvisable. A resulting design package is then publicly advertised, typically inviting bids from any interested, responsible contractor. On bid opening, a contract is awarded to the bidder with the best technically and commercially evaluated bid. This approach is favored in the design and contracting world since it allows open competition and usually results in the best price. This process works well for well-defined and thoroughly designed projects. With this method, the owner retains responsibility for operating the facility and carries any risk that is transferred from construction into operations.

*Characteristics*

- Two contracts (design/engineer and contractor), three phases (design, bidding, construction)
- Best understood by all parties
- Linear sequence of work typically results in a longer delivery cycle.

*Advantages*

- Owner retains control of design (only an advantage if the owner has excellent in-house capability and knowledge of what it requires).
- Procurement laws and processes well defined
- Low cost (through open bidding)

*Disadvantages*

- Owner approves the design, consequently bearing liability for misalignment of plans and specs.
- Potential changes to final cost (owner responsible)
- Most litigious method
- Lack of contractor input during project design and objectives resulting in a lack of collaboration on materials, construction techniques called out in drawings, etc.
- Owner retains all process and long-term operations risk.
- Contractors provide everything to minimum specification and may attempt to cut corners to boost margins.
- Requires high levels of inspection and control by the owner

**Construction management/multiple prime (CM–multi-prime).** CM–multi-prime is a variation of DBB through multi-prime contracting. In this method, the owner bids and holds separate contracts directly with contractors in multiple construction work disciplines (e.g., general construction and civil, structural, mechanical, and electrical engineering). Through this approach, the owner, acting as a general contractor or through a separately contracted construction manager, manages the overall schedule and budget of construction delivery. This variation of the DBB contracting method has gained favor as a method of better construction prices, whereby discrete projects in each construction discipline are bid separately, allowing the owner to award construction contracts on the first portions of the project as soon as the respective aspect of design is completed for each discipline. The owner has more control over the project schedule and budget and sets the timeline for bidding individual portions of the work. Additionally, the owner has the opportunity to achieve savings by directly procuring major material items and avoiding contractor markups.

*Characteristics*

- Potentially multiple contracts (designer, contractor, subcontractors, etc.)
- Linear design of the project
- Construction manager selected based on qualifications

*Primary reason to choose*

- Owner maintains control of design.
- Contractor(s) involved early
- Combined fast track and lowest bids

*Disadvantages*

- Potential changes to final cost for scope changes, overlaps, or gaps
- Little subcontractor input to project design and objectives
- Owner exposure to construction manager's lack of proper oversight and/or significant managing resource input required from the owner
- Owner retains all process and long-term operations risk.
- Owner must have detailed knowledge of construction and constructability to manage this method well.

**Construction management at risk (CMAR).** CMAR is a project delivery system in which an owner contracts with a construction manager based on qualifications, experience, and a guaranteed maximum price (GMP) to manage and build a project. The CMAR provides professional services and acts as consultant to the owner in the design, engineering, and construction phases. The CMAR may also provide the actual construction of the project depending on its expertise, ability, and availability of the firm. The key factor in successful project delivery is that the CMAR must manage and control project costs to not exceed the GMP, as contractually costs exceeding the GMP that are not change orders are financial risk to the CMAR. However, because the CMAR is not receiving the benefit of being paid for the entire project, only for the construction management, it is unlikely that the CMAR will pay for any overrun cost beyond this profit, which may be relatively small.

*Characteristics*

- CM selected based on qualifications and fees
- Some construction risks transferred to general contractor
- Similar to CM multi-prime for selection and management of the work
- Open book on costs (subcontractor and supplier payments) and procurement process
- Higher level of cost control for owner from the beginning of the project
- Flexibility to price the project
- Subcontracts reassigned to the CM
- Bonding possible for the entire scope of the work (GC and subcontractors)
- Risks can push the CM not to act in best interest of the owner.

*Advantages*

- Opportunity to establish selection criteria to match objectives of the project
- Owner control of the final design process
- Early contractor involvement for better understanding of the contract
- Better risk allocation and fewer claims
- Establishes GMP, which provides greater budget certainty
- Establishes partnership among the owner, designer/engineer, and contractor early in process
- Burden of managing and coordinating the project off the owner

*Disadvantages*

- Potential duplication of administration activities
- Less competitive leverage on the general contractor when pricing construction
- Fast-tracking difficult to control
- Major changes in scope will result in change orders; owner responsible for changes
- Owner's qualification-based selection of CM rather than other criteria (i.e., lowest bid)
- Potential for designer/engineer clash with CM during design
- All process and long-term operations risk on owner
- Everything done to minimum specification, corners cut to preserve or boost margins

**DB.** DB is a project delivery method in which a single party performs the design/engineering and construction of a project. The basic concept is that one firm is selected under one contract with project workflow from initial concept and design through completion.

*Characteristics*

- Single point of contact/responsibility
- Integrated team often the fastest delivery method
- Most cost effective
- Effective with a well-defined scope/design
- Need for timely decisions
- Must effectively administer DB process

*Advantages*

- Detailed design control transferred to the DB contractor
- Construction input during the design process
- Overlaps and gaps in scope identified during preconstruction
- Cost–benefit of procuring the construction directly from the trades
- No mark-ups on subcontracts or on changes
- Improved schedule due to early resolution of design and construction issues
- Early construction start possible because of packaging of work
- Tighter control to adhere to scope budget
- Better risk allocation
- Clear project goals
- Reduced delivery time
- Single source of responsibility
- Enhances innovations, partnering
- Early knowledge of project costs, GMP identified
- Integration of design and construction

*Disadvantages*

- Owner responsible for changes, overlaps, and gaps in scope as given to the DB contractor
- Loss of owner control of design aspects not detailed in the contractual documentation

- Possibility that subcontractors may be brought into project late in the process
- Need upfront program and performance criteria
- Owner makes quality-based decisions
- Owner pressured for early decisions
- May require a cultural change
- All process and operations risks the responsibility of the DB contractor until hand-over and then, owner in charge
- Variations of DB:
  - Modified DB: a project delivery method where the owner completes a substantial portion of the design, and a single entity completes the remaining portion of the design and builds the project under a single contract.
    - Advantages:
      - Ability to innovate
      - Single source of responsibility
      - Fewer owner resources required
      - Better project coordination
      - Improved risk management
      - Reduced project delivery time
      - Cost savings
    - Disadvantages:
      - Early contractor input not realized
      - Contractor qualification issues and control of the adequacy of the proposed design

**DBOM.** DBOM is a project delivery method where a single entity performs the design/engineering and construction of a project and operates the resultant project for a specified period of time under a single contract.

*Advantages*

- Integrates design, construction, and operations/maintenance under one single contract
- Operational issues identified and considered during design
- Faster project delivery
- Better life-cycle costs
- Facility performance guarantee provided by DBO firm
- Fewer members of owner's O&M staff required
- Fixed long-term O&M overhead costs and fixed energy consumption
- Process design and operations proved during initial years of operations period (typically 1–3 years)

*Disadvantages*

- Potentially longer procurement process
- Costly to engage a DBOM contractor because of the financial risks taken on by the contractor
- Requires strong advisory support (legal and technical)

**PPP or P3.** Note that references to the private sector entity in the following text refer to the partner in the PPP. It may consist of a consortium, or if an individual entity, that entity will typically subcontract specialized portions of the work such as the intake and outfall design. Further, it is typical for the private sector entity to create a project-specific special purpose company (otherwise known as a special purpose vehicle [SPV]), to which it supplies the project equity, and it is the SPV that assumes any project-specific debt. It is the SPV that signs the contract with the public sector partner, its creation typically being a condition precedent of the contract. The key item in this structure for the public-sector client to note is the closeness of the link between the equity donor, process design, and responsibility for the long-term O&M. It is important for public entities, lenders, and project sponsors to not only focus on the design and construction of the project in their evaluation of bidders but also on the long-term O&M experience. This is a relatively new form of project delivery in the United States even though it is used widely in other countries.

The primary purpose of the PPP or P3 contractual approach is to optimize the sharing of project risk between the parties. It also includes for the private sector partner to invest equity into the project; however, unlike precursors like the BOO and BOOT contract models in which the private sector partner also arranges the project debt, P3s usually take advantage of the low-cost debt available to the public sector. This is particularly true in the United States with access to state revolving funds and federal programs like the Water Infrastructure Finance and Innovation Act program.

A frequent question then for public sector entities is this: Why add on cost to the project via the private sector equity when the project could be fully funded using the available low-cost finance? The question ignores the internal costs of the public-sector entity managing the project itself, including associated consultancy, engineering, legal fees, and any project cost and program overruns, as well as the treatment process design risk and long-term operations risk. It should be noted that cost and program overruns have the effect of increasing the cost of the low-cost finance. The apparent premium to be paid for the equity return can be considered as the cost to transfer the treatment process risk and long-term O&M risk to the private sector. It is this transfer of risk that is particularly suited to seawater desalination projects, given the relative lack of historical experience in the public sector of both the treatment process and plant operations.

The previous paragraph does highlight a frequent misconception of the work required by the public sector partner before starting a P3 contract bidding process. While it is necessary to hire a strong advisory team (legal, financial, and technical) to prepare a conceptual project design and accurate RFQ and RFP documentation based on performance requirements, it is not necessary to fully develop either the design or equipment specifications. This task falls under the responsibility of the private sector partner who will assume the process and long-term performance risk. Therefore, the upfront project development costs for the public sector client can be considerably less than for a traditional DBB or similar form of project development and delivery.

By definition, P3 contracts are long term, typically 15–30 years. The term is dictated by a combination of the term of the long-term project debt and the unit purchase price the public sector partner is willing or able to pay. This is another major difference between P3 contracts and traditional forms of project delivery: The pricing is declared in terms of a unit purchase price (also known as the unit sales price). In the case of a seawater desalination plant, the unit purchase price would typically be based on a fixed monthly payment and a variable monthly payment. The fixed monthly payment cost includes all equity return, as well as legal, design and construction, commissioning, and fixed operations and maintenance costs for the term of the contract. The variable monthly payment is the price per volume of desalinated water contracted for each month and includes the variable O&M costs like chemical usage. Power costs (i.e., electrical consumption) may be omitted from the operations cost because the public sector partner has an existing contract with the power provider or perhaps is the power generator itself. To ensure that the efficiency of the plant is maintained, the contract will typically include a clause defining a maximum power consumption per unit volume produced, with any excess for the account of the private sector partner. There is typically a take-or-pay provision in the contract; i.e., the public sector signs to pay for a minimum (or base) volume for the period of the contract, often with a ceiling volume to be made available for peak demand periods. These are two key parameters to be agreed between the parties at the outset of the project. At the end of the contract term, the ownership of the plant will transfer to the public sector partner in full operating condition and with contractually stipulated equipment spare parts, etc.

Responsibilities of each of the entities in the execution and operation of the project include the following.

*Responsibilities*

- Public entity
  - determines the technical feasibility of the project and defines the minimum performance guarantees (including the minimum and maximum flow production and water quality requirements) and the term of the O&M period;
  - evaluates and assesses the proposal(s) from the private sector;
  - executes a contractual arrangement, i.e., a water sales (or purchase) agreement with the private entity;
  - arranges the project debt (if required); and
  - oversees and monitors the private entity's performance to the program.
- Private entity
  - designs the project;
  - delivers the project either directly or via subcontracts;
  - is responsible for providing equity and any delta to the public sector partners debt arrangement;
  - is responsible for the risks associated with the design, procurement, construction, and O&M of the project;
  - operates, maintains, and manages the resultant project for the contractual term/period; and
  - returns the project to the public entity following completion of the contract period.

*Advantages*

- Expedited completion compared to conventional project delivery methods
- Project cost savings and better life-cycle costs
- Improved quality and system performance from the use of innovative materials and management techniques
- Substitution of private resources and personnel for public resources
- Facility performance guarantees provided by private firm
- Access to new sources of private capital
- Reduction in risk for public entity
- Provision of more opportunities for technical innovation
- Definite price of water for public entity partner, which enables accurate future budgeting

*Disadvantages*

- Major cultural shift to implement and limited experience in the US water industry
- New form of contractual agreement for the public entity partner requires strong advisory support (legal, financial, and technical) to achieve a mutually beneficial project
- Political (non)acceptability
- Potentially longer and more costly procurement process
- Long-term contract a take-or-pay volume and typically functions as a liability for the public-sector partner.

**Alliance method.** One final alternative delivery procurement strategy that has been successfully employed to manage all the risks and variables of complex desalination project delivery is the alliance method. An alliance is a delivery mechanism that differs markedly from traditional project contract models. An alliance occurs when two or more entities come together under an agreement and agree to jointly achieve a specific outcome whilst sharing all the risks and rewards associated with achieving the outcome. It may be an operational or capital project outcome. There may be one or more owners and one or more service providers (e.g., designers, contractors, process providers, and other suppliers).

Alliances have a number of benefits over traditional contracting models. However, one of the basic tenets of alliance methodology is that the ultimate end-user or owner of the project pays the real cost of the project. When there are many unknown risks and variables, this is a reasonable position to adopt. The financial model in an alliance agreement, however, strongly incentivizes creativity and innovation to drive down costs and improve project outcomes. It also has powerful disincentives associated with cost-overruns and poor project outcomes. In this manner, all parties to an alliance agreement strive to drive down costs and deliver exceptional project outcomes as all parties receive benefits from achieving such outcomes.

The alliance method is not necessarily the right procurement strategy for every project. A project ideally will have several specific features that will lead to the conclusion that an alliance is the best choice of delivery. One feature is the risk profile of the project. Seawater desalination plants have an uncommon risk profile. Power supply, proximity to water distribution system, seawater intake, concentrate discharge requirements, and other environmental considerations result in limited siting opportunities. Energy recovery, pump life, pretreatment options, other treatment processes, and operating risks associated with the technical complexity and energy consumption of a desalination plant are generally somewhat greater than those of conventional water treatment plants. These factors may be well managed using an alliance method.

It should be noted that there are many similarities between the P3 and alliance approaches. However, the selection of an alliance approach rather than a P3 might be the preferred portion when uncertainty whether the project will actually continue to completion and the client does not want to commit to a long-term take-or-pay agreement.

A positive example of the alliance procurement method to develop a seawater desalination project has been demonstrated in Australia. In Western Australia, the Water Corporation supplies water services to approximately 2 million people. A drying climate since 1975 created significant challenges for water supply and required establishment of permanent water restrictions.

The Water Corporation as project owner developed an alliance agreement under which they, with a designer, a process provider, and a contractor, would agree to deliver the Perth Seawater Desalination Project jointly. Under the alliance agreement, all parties assumed collective responsibility for sharing all project risks and rewards.

A lot of work still goes into setting up an alliance. The client organization must deal with site selection, power supply, integration into the water supply system, technical specification for project deliverables, environmental permitting, stakeholder engagement, and certain decisions related to desired technology (e.g., type of energy recovery). Consideration must be given to whether there will be an operating contract and how to transfer construction risk into an operations contract and many more such things.

Traditional contracts distribute risks and responsibilities to different parties, and there are prescribed legal or commercial consequences if the parties do not meet the requirements of the contract and its risk allocation. By contrast, in an alliance, all parties assume collective responsibility for risk management. They all share in the commercial consequences of poor management, or alternatively, they all share in the benefits of good mitigation and management. Pain or gain is measured and remitted against predetermined targets that have been jointly agreed upon.

There are different methods available to set up an alliance. The method used will depend on the client organization's specific governance and other issues and risks that they wish to address through the establishment method.

## 6.3 PROJECT DELIVERY SELECTION APPROACH

The owner's engineer should identify the potential project delivery methods for the project in coordination with the public sector entity and then tabulate the advantages and disadvantages for each in a matrix format. Since there is no single "best delivery method" suitable for all projects, the most appropriate delivery method(s) for each project should be selected based on the comparative weighted analysis of the features, benefits, and risks for achieving the project objectives. The engineer's approach should be to create a shortlist from the delivery methods by considering, at a minimum, the following project characteristics:

- complexity and interfacing issues;
- the owner's desired level of control and experience on similar projects;
- the owner's resources;
- local regulation requirements;
- key project objectives, including schedule, cost, and scope;
- identification and allocation of risks;
- opportunity to introduce innovative technology; and
- funding and financing.

The engineer should then primarily assess the project against the above or similar characteristics using a scoring matrix with weighting factors applied to each characteristic. The normalized score of each delivery method can then be plotted on a predetermined suitability scale (essentially divided into categories) with the traditional contract (with a higher public responsibility and level of control) at the lower end, followed by design and construction, management, relationship, and finance methods (with a greater private responsibility) at the higher score end of the suitability scale (recall these method mentioned earlier in this chapter). Essentially, the engineer should move within the "relationship spectrum" from adversarial to nonadversarial delivery-method-relationships by matters of degree. The process is designed in such a way that based on the above characteristics, the higher the score, the less adversarial the delivery method becomes. In general, building projects of standard design and/or a well-defined brief may be better suited to a procurement method at the lower end of the scale (higher public responsibility/adversarial). On the other hand, complex projects requiring innovation may be better delivered using approaches that are more toward the higher end of the scale (greater private responsibility/nonadversarial). One could envision the selection of two to four (maximum) delivery methods by the end of the first exercise. In addition, it may be necessary to conduct a market-sounding exercise and receive feedback from potentially interested parties before moving forward with a selected alternative.

The public entity, guided by the engineer, should then consider the use of a multicriteria analysis method (and if necessary, a facilitated project stakeholder workshop) to further evaluate the short-listed delivery methods. The evaluation criteria should be based on aspects such as project objectives (e.g., certainty of delivery by target date) and risk treatment measures (e.g., increased operator engagement, flexibility to cope with variable quantities).

When developing the project delivery services, the project management professional should incorporate all requirements of the project scope. Critical aspects and potential impacts determined from this schedule should then be incorporated into the evaluation criteria. Weightings are then assigned to the criteria based on their relative importance to the project. Each delivery method is then scored for its performance against the evaluation criteria as listed in the multicriteria analysis table produced. The analysis can be strengthened by project-specific aspects like special conditions of contract and the use of incentives.

The delivery methods with the highest weighted scores should then be selected for the project and ranked. As a final check, the preferred delivery methods should be reviewed to identify any weaknesses, such as low score of specific criteria, in the multicriteria analysis.

The final project delivery plan is then prepared, incorporating those aspects as outlined, the scope of the work, and any additional aspects that may be deemed applicable. An assessment of time and budget needed for contractor procurement is then produced for the recommended delivery method(s).

## 6.4 PROCUREMENT METHODS BREAKDOWN

Alternative procurement methods such as BOT, BOO, and DBFOT have become standard procurement methods for seawater desalination projects outside of the United States since they allow the transfer of most of the risks to the private entity, allow for innovation and optimization of life-cycle cost, and do not require the owner to find the upfront capital investment to implement the project. It shall be noted that another procurement approach widely used in the Middle East, where demand for both electricity and saltwater desalination is high, is called the integrated water and power project (IWPP). IWPPs consist of a combined facility serving both as a desalination plant and a power plant. In a minority of markets, the DBB approach is used. The EPC approach is used by the entity that prefers to remain the owner of the asset and operate the plant themselves.

## 6.5 SUMMARY

There are several well-established procurement strategies and mechanisms to deliver a seawater desalination plant project. Selecting the most appropriate project procurement alternative is crucial to delivering a successful project, which will depend on local regulation requirements, permitting, technical requirements, and the owner's financial capability. Key project drivers like risk sharing, level of control, cost, schedule, innovation and experience, project financing and funding, conditions of the construction market, and the economy also impact success. Seawater desalination projects are considered high-risk projects, and to minimize the life-cycle costs, the majority of these projects developed around the globe within the last several decades are implemented using alternative construction and procurement models. It is important for the public entity to evaluate the need for an advisory support team early on for assistance in selecting the best procurement strategy for the project, establishing an effective and efficient procurement process, helping the public entity understand its risk exposure during the preparation of the procurement documents, and providing advice and support during contract negotiations. When considering the implementation of an alternative delivery approach, it is also recommended that the entity prepare a project implementation plan that details the scope of work and organizational and resource needs because an alternative delivery approach requires very different skills and resources than the traditional DBB implementation.

Chapter 7

# Plant Operations and Maintenance

*Second Edition Authors: Say Sengsouvanna, Keith Downer, Greg Wetterau, Mo Malki, Rajendra Vaidya*
*First Edition Authors: Gary Silverman and Irving Moch*

## 7.1 INTRODUCTION

Effective desalination plant long-term performance depends on good operations and maintenance (O&M) of the component systems. Successful O&M strive to optimize system performance, maximize useful life of the equipment, promote regulatory compliance, and minimize costs. This includes proper staffing, process monitoring, cleaning, routine and preventive maintenance, inventory management, training, and record keeping and reporting. Each desalination plant has its own specific requirements and management, so O&M can be accomplished in many ways. This chapter is not an O&M manual for seawater desalination plants; each plant will require its own specific manual. It does, however, make the reader aware of the key components that need to be addressed to allow for the optimal life-cycle cost of any facility and draws attention to the differences between the O&M of a traditional water treatment plant and those of a seawater desalination plant. Note that seawater desalination plants usually operate at pressures around 1,000 psi (6,900 kPa), a significant difference from the typical gravity-functioning traditional plants, thereby requiring a robust health and safety protocol to work in conjunction with the O&M methodology.

The key performance indicators (KPIs) for the seawater reverse osmosis (SWRO) plant O&M include

- meeting product water quality and quantity requirements,
- frequency of membrane replacement,
- frequency of cartridge filter replacement,

- power consumption and energy recovery,
- frequency of membrane cleaning,
- chemical costs and/or normalized chemical consumption,
- waste disposal costs, and
- maintenance costs.

A common measurement of plant performance is plant availability or the percentage of time the plant's production was at or above its design capacity, quantity, and quality. It is not uncommon for desalination plants to have availabilities above 95 percent with exceptional plants having availabilities as high as 98 percent. High desalination plant availability cannot be achieved without a robust O&M program and well-trained and experienced staff. Some desalination plants are planned, designed, and operated for seasonal demands, thus complicating the defining plant availability. The other KPIs of power consumption, chemical, and maintenance costs have a wide range of values depending on treatment technologies, feed water conditions, cost of power, and chemical costs. These relative costs can be minimized with proper O&M with the understanding that trade-offs can exist. A typical trade-off is reducing long-term fouling risk by more aggressive cleaning of membranes at longer durations versus less aggressive cleaning at higher frequency and with shorter duration between cleanings.

## 7.2 MANAGEMENT AND STAFFING

Desalination plant staffing and management structure should reflect the facility capacity, process complexity, level of automation, legal requirements, the owner's staffing policy, and the owner's execution strategy—including whether to contract out services such as O&M or, perhaps forced by limited readily available companies with the necessary expertise, whether to carry out O&M by direct hire. Besides the management and administration necessary for any plant, the desalination process involves several technologies that require specific skills and training to operate and maintain. It is vital that the personnel have sufficient knowledge and training for appropriateness and adequacy of the pretreatment and high-pressure membrane process control, chemical cleanings, and equipment repair and maintenance. Modern desalination plants require skilled mechanics, pipefitters, electricians, technicians, and increasingly, programmers, due to the extensive use of automated control systems used at these facilities. Almost certainly some aspects of the maintenance will have to be contracted out to individuals or companies with specialized expertise.

The number of staff typically required to operate and maintain equipment increases with increasing water production capacity, treatment process stages, or number of work shifts. To reduce the staffing requirements, desalination plants today are commonly designed with a high level of automation. Some jurisdictions may have their own minimum staffing requirements for desalination plants, including security, especially if the desalination plant is considered part of the critical infrastructure. The last factor is related to the owner's staffing policy as well as the nature of the facility. For example, some owners may prefer to always have an operator(s) at the plant, while others may prefer smaller operations staff and leaving the plant unattended at times. The owner's staffing policy may require additional staff to meet the company's health and safety policies, as is common for industrial desalination plants, particularly in the mining industry.

### 7.2.1 O&M Execution Strategy–Owner

Desalination plants are normally operated in two different ways: owner operation and contract operation. Under the owner-operator model, the plant owner operates and maintains the facility with its own staff. With the owner-operator model, the owner bears all the responsibility for production, maintenance, staffing, training, inventory management, and regulatory compliance. The major benefit of the owner-operation model is that the owner maintains the highest level of control over costs and production. Small industrial desalination facilities producing water for boiler makeup, semiconductor, or pharmaceutical industries typically use the owner-operator model.

Alternatively, the owner may employ another company that specializes in O&M of desalination facilities to run the plant for a fixed time and price. These contract operations can include both O&M or only maintenance while the owner maintains the operations responsibility. O&M contracts may also include provisions for cost sharing on power or chemicals, minimum plant investments, performance or production guarantees, and owner personnel training with the intention that the owner can assume operations in the future. The major benefits of contract operations are access to the company's specialized services, experienced personnel, and fixed costs over the contract term. Medium and large desalination plants in the municipal water sector, mining, or other sectors where water production is not core to the industry increasingly favor the use of the contract operations model.

An emerging model for the O&M provision in the municipal sector is the delivery of the desalination facility using the public–private partnership (P3) project delivery contract, in which the specialist company not only takes responsibility for the O&M of the facility but also the process design, project delivery, and subsequent long-term compliance. This eliminates any future conflict over O&M issues with third-party design and construction firms and perhaps most importantly, ensures that sufficient funds are available to efficiently and effectively maintain the facility. The principal reason for municipal-owned seawater desalination plants to operate suboptimally (in terms of on-line availability and actual production volume) is the lack of budget available for both replacement parts and training and retaining staff.

### 7.2.2 Typical Staffing Requirements

The actual staffing of a desalination plant varies, but a common staff structure is composed of at least one plant manager with operations staff (shift supervisors, control room operators, field operators, and laboratory analysts) and maintenance staff (mechanical, electrical, programmers, and instrumentation and controls technicians). For smaller facilities, cross-discipline staffing is more practicable, where both O&M roles are filled by all staff members. Large plants are often staffed with additional personnel such as a dedicated operations manager, maintenance manager, health and safety staff, and maintenance discipline supervisors. They also typically employ a process engineer to manage the complex interlocking components comprising the pretreatment, desalination, and posttreatment processes.

In addition, in some jurisdictions (e.g., the public water sector), desalination plants will be required to be operated under the supervision of licensed personnel with minimum experience requirements dependent on plant capacity and treatment processes being used. The type of license, number of license holders, authority issuing the license, and license cost and upkeep all vary from jurisdiction to jurisdiction. Table 7-1 provides an estimate of typical plant staffing for desalination facilities of varying capacity based on a desalination planning document produced by the US Department of the Interior (US Bureau of Reclamation 2003). SWRO facilities typically require more staffing than brackish water facilities, due to the more complex processes (including pretreatment needs), higher operating pressures, and more frequent chemical cleanings.

**Table 7-1 Desalination operations and maintenance (O&M) staff**

| Desalination Process | Plant Capacity *million gal/day (million L/day)* | | | | |
|---|---|---|---|---|---|
| | 1.0 (3.8) | 5.0 (19) | 10 (38) | 25 (95) | 50 (189) |
| Thermal evaporation | 14 | 14 | 18 | 20 | 35 |
| Seawater reverse osmosis | 6 | 12 | 15 | 18 | 20 |
| Brackish water reverse osmosis | 6 | 10 | 12 | 15 | 17 |
| Electrodialysis/electrodialysis reversal | 6 | 10 | 12 | 15 | 17 |

*Adapted from* US Bureau of Reclamation 2003

## 7.3 PROCESS MONITORING AND CONTROL

The performance of an SWRO system depends on the control of the pretreatment and membrane systems. Monitoring system parameters like feedwater quality, pretreatment performance, temperature, membrane pressures, and permeate quality are all essential to identifying problems early on, so they may be resolved more easily. There is an increasing need to integrate climate and regional source water modeling due to the high variability in source water of desalination plants with high impact events as algal blooms, storms, and wind patterns. Proper plant monitoring and control is key to maximizing membrane life and plant production while reducing operating costs and unplanned maintenance. Knowledge of nominal process values is integral to identifying and addressing unusual operating conditions quickly to prevent equipment failures and maintain maximum availability.

### 7.3.1 Process Monitoring Baseline

Process monitoring and control begins with plant commissioning and startup. Along with identifying problems with the construction and installation of the equipment, the first few days during and after commissioning is when initial configurations of membranes, physiochemical makeup of the raw water, initial pressures, flows, and water quality are recorded determining the baseline performance of the plant in the new state. At a minimum, after the initial 24–48 h of plant operation, all process parameters should be recorded for an additional 48 h to establish the baseline record of performance. Beyond commissioning and startup, the plant or portions of the plant should undergo a "re-baseline" after a major equipment or membrane change occurs. The baseline record will become part of the basis of the ongoing plant monitoring, control, and maintenance decisions. Adopting a thorough performance specification from the membrane manufacturer and system integrator will assist plant owners and or operations to successfully decide on the level and frequency of process monitoring parameters and associated actions to document operational performance.

### 7.3.2 Process Monitoring

While not exhaustive, Table 7-2 lists typical monitoring parameters, their frequencies, and signs of potential problems for desalination plants.

Most modern SWRO plants have extensive automation, including a variety of online instrumentation for the monitoring of process parameters. These systems typically include alarms to notify operators of abnormal conditions in the plant that require their action to correct. Even with advanced automation systems in place, it is important to periodically verify process parameters using handheld instrumentation and laboratory analysis.

**Table 7-2 Process parameter monitoring for seawater reverse osmosis (SWRO) desalination plants**

| Parameter | Typical Frequency | Potential Indicator for |
|---|---|---|
| **Feedwater Monitoring** | | |
| **Parameter** | **Typical Frequency** | **Potential Indicator for** |
| Feedwater quality -pH, temperature, turbidity, conductivity, oxidation-reduction potential, silt density index (SDI), algae, microbiology, biostability, etc. | Continuous to annually, depending on measurement | Scaling, biofouling potential, algae blooms, petroleum presence |
| Feedwater pressure, well level, or vault liquid level | Daily | Biofouling of plant intake structure/ pipeline, biofouling of pump suction or discharge piping, biofouling of screening equipment, well degradation |
| **Pretreatment Monitoring** | | |
| **Parameter** | **Typical Frequency** | **Potential Indicator for** |
| Filter differential pressure or transmembrane pressure | Continuous | Feedwater quality change, ineffective backwashing |
| Backwash frequency | Daily | Feedwater quality change, ineffective backwashing |
| Filtered water quality (turbidity, total organic carbon, 15-min SDI) | Continuous to daily | Hydraulic rates too high, suboptimum coagulation, pretreatment not optimized, membrane integrity compromised (if microfiltration/ ultrafiltration) |
| Chlorine residual, oxidation-reduction potential | Continuous | Risk of membrane damage due to oxidation |
| **Seawater Reverse Osmosis Monitoring** | | |
| **Parameter** | **Typical Frequency** | **Potential Indicator for** |
| Feedwater conductivity, temperature | Continuous | — |
| Flows (permeate, concentrate) | Continuous | Membrane fouling |
| Pressures (feed, concentrate, permeate) | Continuous | Membrane fouling |
| Differential pressure drop across the pressure vessel | Continuous | Colloidal fouling, biofouling |
| Permeate conductivity | Continuous | Membrane fouling |
| Pressure vessel conductivity profile | As required | Membrane abrasion<br>Brine seal rolled<br>O-ring problems<br>Membrane oxidation or fouling |
| Concentrate conductivity | Continuous | Membrane fouling |
| Membrane salt rejection | Continuous | Scaling |
| High-pressure pump temperatures | Continuous | Bearing replacement, motor failure |
| High-pressure pump efficiency | Monthly | Impeller wear, bearing replacement |
| Energy recovery device efficiency | Monthly | Varies |
| Membrane autopsy | As required | Foulants, membrane age, oxidation, mechanical damage |

*(continued)*

**Table 7-2 Process parameter monitoring for seawater reverse osmosis (SWRO) desalination plants (*Continued*)**

| General Monitoring | | |
|---|---|---|
| **Parameter** | **Typical Frequency** | **Potential Indicator for** |
| Product water quality | Continuous | Out of specification |
| Chemical dosing calibration | Weekly | Excessive chemical costs |
| Bulk chemical concentration | On delivery and monthly for liquids that degas (e.g., sodium hypochlorite) | Chemical degradation<br>Incorrect chemical supply<br>Improper dilution |
| Membrane cleaning frequency | As completed | Ineffective cleanings<br>Pretreatment problems |
| Membrane cleaning performance | As completed | Ineffective cleanings |
| Concentrate quality | As required by permit | Environmental toxicity |

In the analysis of the monitored process parameters, it is important to understand that parameters of permeate flow and salt passage are affected by each other and several other factors such as temperature, recovery, and raw water salinity. When reviewing these parameters for abnormal patterns, it is important to normalize the data to eliminate the effects of these other parameters on the data.

## 7.3.3 Standard Operating Procedures

Standard operating procedures (SOPs) are critical to ensure all operators at any given SWRO plant consistently, correctly, and safely follow the best practices for the plant. SOPs describe in step-by-step detail routine or critical plant control operations. Critical plant control operations are routine or expected tasks that, if done incorrectly, could cause lost availability or a safety incident. SOPs should be concise and unambiguous, ensuring precise communication of instructions for performing the tasks, improving execution, maintaining quality, increasing safety, and providing consistent implementation necessary for efficient planning. They should be prepared by the appropriate subject matter expert in collaboration with relevant stakeholders. For example, a clean-in-place (CIP) SOP might be developed by the person in charge of commissioning the CIP system. The responsible person would need to consult with the CIP process designer and operators to write a succinct step-by-step procedure that communicates all the required steps to perform the task as intended.

SOPs are plant-specific live documents that require regular updating as the best practices for the facility are refined. Independent review by one or more individuals experienced with or knowledgeable about the procedure should be done before approval of SOPs. Each SOP should contain a revision log that documents the original author, reviewer(s), and dates of reviews and approvals; if changes are made to a procedure, the SOP should be updated and reviewed again, and the log should be revised accordingly. Table 7-3 lists some of the core SOPs that a typical SWRO desalination plant should have.

## 7.3.4 Laboratory Control

Implementing and maintaining laboratory quality control (QC) practices is necessary to ensure that analytical methods are performed correctly and consistently and that results are both accurate and precise. A well-run QC program supports operations by providing confidence in water quality analytical data used for evaluating treatment effectiveness, allowing engineers and operators to tune system performance as needed to achieve operational goals.

**Table 7-3 Typical standard operating procedures (SOPs) for seawater reverse osmosis (SWRO) desalination plants**

| Process | Sequence Topic |
|---|---|
| Plant-wide | Plant startup and shutdown |
| | Daily routine parameter checks |
| | Lockout/tagout procedure |
| Pretreatment | Silt density index sampling and monitoring |
| | Media filter backwashing sequence |
| | Media filter conditions inspection |
| | Membrane fiber repair (for microfiltration and ultrafiltration) |
| | Cartridge filter replacement |
| Membrane process | Individual train startup and shutdown |
| | Membrane preservation and layup, short- and long-term storage |
| | Membrane loading/unloading |
| | Vessel probing and profiling |
| | Clean-in-place processes |
| | Membrane flushing |
| Chemical systems | Chemical bulk loading and unloading |
| | Chemical batch and solution makeup |
| | Chemical metering system calibration |
| | Chemical neutralization and disposal |
| | Chemical spill cleanup |
| | Chemical spill emergency response |
| Laboratory and instrumentation | Equipment calibration and certification |
| | Instrument calibration |
| | Chemical waste disposal |
| | Jar testing |

Initial QC practices include

- providing training and verifying capabilities of personnel to take water samples correctly and perform analytical methods,
- determining method detection levels for methods and analytical equipment being used, and
- determining operational ranges (upper and lower limits) for methods and analytical equipment being used.

Ongoing QC practices include

- calibrating methods or standardizing titration reagents using SOPs (e.g., consulting *Standard Methods for the Examination of Water and Wastewater*);
- periodically verifying calibration by analyzing a calibration standard or blank and taking corrective action if calibration verification fails (i.e., results exceed ±10 percent of true value) and confirming results with a reanalysis;
- calibrating laboratory instruments (analytical balances, pH/ conductivity probes, turbidimeters, etc.); and

- periodically verifying standardized titration reagents, once per month or sooner, if there is reason to believe that the normality of the reagent has changed (NOTE: Standardized titration reagents are typically stable for several months when sealed and stored properly; if the titration reagent's normality has changed, the solution should be discarded, and a fresh titration reagent prepared and standardized).

## 7.4 MEMBRANE CLEANING

Reverse osmosis (RO) membranes used for seawater desalination should be periodically cleaned using chemicals or elevated temperatures to remove deposits (e.g., foulants, scalants) from membrane surfaces and channels. CIP or in situ cleanings are conducted to restore operational parameters (e.g., permeability, differential pressure, transmembrane pressure) to their baseline values by targeting removal of suspected contaminants. CIP cleanings are typically monitored by operators via a supervisory control and data acquisition (SCADA) system (Figure 7-1).

It is generally better to clean earlier rather than later to minimize chemical concentrations, increase cleaning efficacy, and prevent irreversible fouling or scaling. All of these factors will extend membrane life and contribute to maximizing plant efficiency; however, frequent or aggressive cleaning can also decrease membrane life. There is therefore a tradeoff between the benefits of cleaning (improved water quality and reduced feed pressures) and the drawbacks (higher chemical use and more frequent membrane replacement). An increase in normalized feed pressure or decrease in product quality is a common sign that cleaning is required. The membrane type, the feedwater quality, and system operating conditions will influence the mode and frequency of cleanings. When cleaning becomes too frequent or water quality targets can no longer be met, the membranes will typically require replacement. Typical membrane life has been 7–10 years; however, some facilities replace membranes as frequently as every 5 years.

Some facilities employ hollow-fiber ultrafiltration (UF) membranes as pretreatment for the desalination membranes. UF membranes often employ regular maintenance cleans or chemically enhanced backwashes (CEBs) using moderate chemical concentrations and less frequent recovery or CIP cleanings using more aggressive chemical concentrations

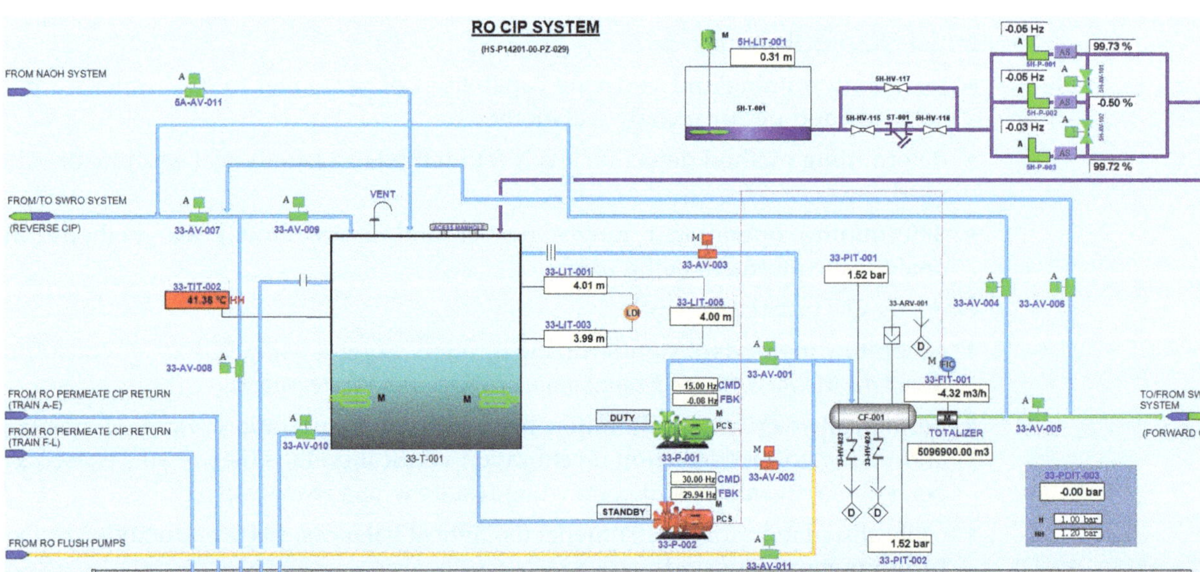


*Source:* Iv3 Aqua Corporation

**Figure 7-1 Screen shot of a reverse osmosis (RO) clean-in-place system SCADA screen**

over longer durations. Maintenance cleans or CEBs are usually scheduled daily or several times a week, while recovery CIP cleanings are done approximately monthly. High-pH cleans are used to remove organic and inorganic foulants and typically use sodium hydroxide and sodium hypochlorite; low-pH cleans are used to remove scale and typically use an organic acid (e.g., citric acid) and sometimes a mineral acid (e.g., hydrochloric acid) for additional depression of pH. Note: Switching to a predominant use of mineral acid versus organic acid can reduce biochemical oxygen demand and chemical oxygen demand loadings to the waste stream and can be a consideration for plant design and operations.

In general, M46, *Reverse Osmosis and Nanofiltration* (AWWA 2007), lists conditions under which a CIP might be considered for RO or nanofiltration (NF) systems:

- an increase in normalized salt passage,
- an increase (10–15 percent) in feed-concentrate (stage) pressure drop,
- a decrease (10–15 percent) in normalized permeate flow rate, and/or
- an increase in normalized feed pressure.

The manual also provides a general cleaning procedure and summarizes cleaning approaches for various types of foulants and scale. Cleaning strategies should always be verified for compatibility with equipment and membrane manufacturers' specifications to avoid damage or loss of warranty. CIPs should only be conducted by operators who are trained in the modes being utilized, understand the safety data sheets (SDS) for the chemicals being used, and have access to the appropriate personal protective equipment.

Frequency of cleaning for RO or NF membranes in desalination is dependent on several factors, including

- ambient water quality (fouling and scaling potential, temperature);
- pretreatment efficacy;
- membrane percent recovery and flux rate; and
- membrane train uptime.

Warmer source waters high in dissolved organic compounds, suspended solids, and sparingly soluble salts will typically have higher biofouling rates that generally increase the frequency of cleaning. If the pretreatment design or operation is inadequate, challenging feedwaters can require RO or NF membranes to be cleaned as often as every month. Limiting the membrane recovery and flux rate will reduce the rate of fouling or

**Table 7-4 Impact of process variable on clean-in-place (CIP) frequency and possible actions to mitigate**

| Process Variable Affecting Frequency of CIP | Possible Impacts to CIP Frequency | Considerations |
|---|---|---|
| Warmer feed water | Increase due to more biofouling | Biofilm monitoring<br>Sanitization following CIP<br>Decreasing the flux rate |
| Feed water high in dissolved organic content | Increase due to more organic fouling | Additional pretreatment<br>Decreasing the flux rate<br>Decreasing recovery |
| Feed water high in concentration of sparingly soluble salts | Increase due to scaling | Additional pretreatment (softening, antiscalant)<br>Reduce membrane recovery |
| Feed water high in colloidal and particulate matter | Increase in feed-concentrate (stage) pressure drop | Additional pretreatment |
| Extended periods of downtime | Increase due to more biofouling | Flushing trains regularly using permeate<br>Laying up the train with a preservative or bioinhibitor |

scaling. Good feedwater quality, pretreatment design, and operational practices can extend cleaning intervals out as far as semiannually. It is preferable to maintain a high uptime for membrane trains to keep the water in the membranes fresh and to limit biofouling. Table 7-4 summarizes the variables that may affect cleaning frequency, their potential impact, and strategies to consider in addressing challenges. Generally, membrane manufacturers will offer membrane warranties based on feed and product water limitations at a CEB or CIP frequency specified as desired by application and or recommended by the manufacturer.

### 7.4.1 Membrane Autopsy

Autopsies can identify specific foulants and scalants on the membrane surface and can determine if chemical or mechanical damage has occurred during operation. The data and conclusions from an autopsy can, in most cases, help identify the cause of a particular problem, which can provide solutions to decrease or eliminate operational issues.

A membrane autopsy consists of various analytical tests, each of which provides specific scientific data to identify and quantify fouling/scaling or chemical/mechanical damage. The results of the different tests are then compiled by experienced chemists and engineers to draw a conclusion.

Membrane autopsies are essential to every SWRO membrane plant. They can be performed on SWRO membranes or on low-pressure membranes such as microfiltration or UF used for SWRO pretreatment. Performing autopsies on pretreatment cartridge filters or even silt density index filters can also provide very useful information about contaminants entering the membrane system.

*Source:* Amaya Solutions, Inc.

**Figure 7-2 Seawater reverse osmosis (SWRO) membrane heavily fouled with suspended solids as a result of undersized upstream multimedia filters**

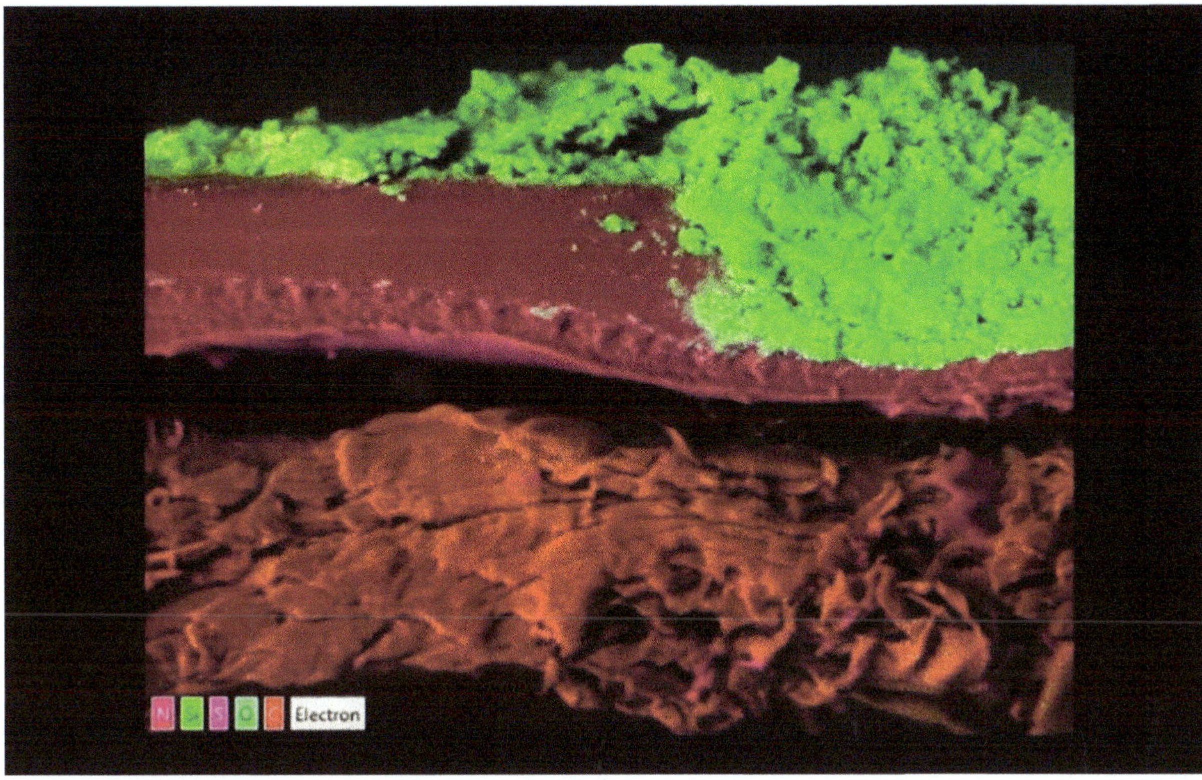


*Source:* Amaya Solutions, Inc.

**Figure 7-3 Scanning electron microscope image of foulants (in green) on a reverse osmosis (RO) membrane**

If not controlled, fouling and scaling will lead to higher operational costs that could cause higher energy demand, increased frequency of cleanings, and reduced lifetime of the membrane elements and in worst cases, increased passage of salts and microbes into the permeate stream (see examples in Figures 7-2 and 7-3). Chemical attack and physical trauma to the membrane surface result in irreversible loss of performance. Identifying the problem at an early stage can help save millions of dollars in membranes before excessive damage renders them useless.

Many sites use membrane autopsies as part of their routine system maintenance program. This provides the advantage of identifying potential problems before they become unmanageable. However, in most cases, a membrane autopsy is only performed when a membrane is failing, underperforming, or suspected of either or when there is evidence of loss of normal performance. A noticeable loss in flow, permeate quality, or visible fouling on the membrane itself would show a problem with the element and the need to have it inspected.

In SWRO system autopsies, it is strongly recommended that one membrane be removed from the lead position and one from the tail (or lag) position. This is important for several reasons.

- The lead element in the system will see the highest concentration of suspended solids.
- The last membrane in the system receives water that has been concentrated and is the most likely to foul from transparent exopolymer particles (TEP). TEP and other natural organic matter (NOM) are typically weak acids that carry a negative charge. High calcium concentrations and high ionic strength cause NOM to form a gel on the membrane surface through calcium bridging and charge shielding, leading to coiling and aggregation.

- Biofouling and NOM (such as TEP, humic acids) can be found on all membranes in the system. However, bacteria that are hydrophobic (repelling or not mixable with water) are attracted to the membrane surface, which is also slightly hydrophobic. Biofouling is therefore always worst at the first stage where the bacteria attach to the first available hydrophobic surface that they find. By contrast, organic acids are hydrophilic (tending to mix with or dissolve in water) at nonacid pH levels. They can therefore concentrate and cause heavier fouling at the tail end.
- Suspended solids and precipitated metals (such as iron or aluminum from coagulant carryover) can also end up on any membrane in the system but will concentrate and cause membrane abrasion in the tail elements.

An autopsy report assembles all the various tests that have been performed. Detailed photographic evidence of membrane conditions is included, as well as extensive documentation of the results of the various autopsy procedures. A good membrane autopsy report can accurately identify specific types of suspended solids, organics, microorganisms, and different inorganic foulants. The autopsy report should be holistic, considering not only findings from the different analyses but also the history of the plant, the applied pretreatment, and the nature of the water chemistry.

As an example, aluminum silicate scales are commonly found in membrane systems using aluminum coagulants and containing high silica in the feed water. But the finding of aluminum silicate on a SWRO membrane would show silts or clays; silica in seawater is barely detectable and does not reach sufficient saturations at the lower recoveries at which SWRO plants operate.

Scaling is very rare in SWRO systems because of the low recoveries at which they operate and the high ionic strength of the water; spectator ions such as sodium and chloride lower the activity of the scale, forming divalent ions through charge shielding. However, there have been cases of severe calcium sulfate scaling attributed to operational errors or lack of maintenance. One common cause is a leaking concentrate check valve causing an offline train to produce permeate from the pressurized brine of the operating trains (the low volume at high pressure results in 100-percent recovery). Other causes may include malfunctioning modulating concentrate valves, miscalibrated flowmeters, or operator error. In all cases, it is essential that all operational information be shared when requesting a membrane autopsy.

Unexplained losses in permeability that occur in the absence of a foulant can often be attributed to changes in membrane chemistry. The chemist or engineer preparing the membrane autopsy report must be made fully aware of all chemicals introduced upstream of the RO in order to identify the cause and effect. Iodine in the feed water has been associated with irreversible losses in permeability. Mild exposures to chlorine in slightly acidic conditions can lower the polarity of the polyamide membrane, resulting in irreversible losses in permeability.

A membrane autopsy should always include a conclusion that helps the plant owner/operator identify the cause of the problems and provide potential solutions for their mitigation. These may include recommendations for changes to the cleaning protocol or alterations to the existing pretreatment program.

The autopsy report is an invaluable tool, whether it is performed for routine maintenance or to determine the cause of an ongoing operational problem.

### 7.4.2 Membrane Relocation and Replacement

As the SWRO plant's membranes age, membrane cleanings become less effective in restoring baseline operating conditions. In some cases, such as when cleanings are delayed or feedwater conditions suddenly affect the system, more direct intervention in membrane performance may need to be taken. Depending on the plant system size and manpower, some systems may attempt to rotate the membranes in the vessels to restore performance before resorting to membrane replacement.

Membrane relocation usually focuses on differential pressure recovery but can target poorly performing membranes specifically if a vessel profile shows that a benefit may be realized. In principle, membrane relocation exploits the fact that the lead two membranes of the vessels experience significantly higher fouling than the tail two membrane elements. The first two membranes experience nearly 100 percent of the feed flow, while the last two are typically only 60 percent. If the first two membranes were moved to the tail position because of reduced flow through those membranes, their contribution to the overall differential pressure would be reduced because of reduced flow in their new position. To further improve the relocation of the lead two membranes, their orientation may be reversed by moving their brine seals and placing them in the tail position in the opposite flow direction. The former tail position membranes can then be placed in the lead position, which, if scaled, may allow them to better recover during the next CIP.

Besides relocations described previously, there are other membrane relocations that can also be employed, as shown in Figure 7-4.

A vessel profile should be done before planning or executing a membrane relocation. Not all membranes allow for the relocation of flow direction, and if the SWRO systems employ an interstage design whereby two different models of RO membranes are installed in the same pressure vessel, membrane relocation should not be employed. A CIP cleaning is recommended before attempting any membrane relocation to avoid transferring debris downstream and to ensure an accurate pre–post comparison can be made. Some membrane relocation methods require significantly more time to accomplish than others, and this extra time consideration should be considered. Membrane relocation is not a replacement for CIP but rather a rare and carefully planned intervention to help improve performance and cleanings.

Eventually cleaning and/or rotation will not be enough to maintain system performance, and membrane replacement will need to occur. Since membrane replacement is a time-consuming labor endeavor and costly activity, it should be planned to best fit the plant's needs. In general, two types of membrane replacements are used: wholesale train replacement and periodic regular replacement.

Wholesale train replacement is the total replacement of all membranes in the most poorly performing SWRO train. The regular conductivity membrane profiles then help select which old membranes from that train can be repurposed in other trains to improve their performance at no additional membrane cost. This wholesale train replacement is then repeated until all membranes have been replaced plant wide. The advantage of this type of replacement schedule is it reduces the amount of train downtime due to membrane replacement and is easier to organize. The major disadvantage of this methodology is that it produces large step changes in plant performance.

With periodic regular replacement, membranes are replaced in all or a few of the trains at a time, specifically targeting poorly performing individual membranes or vessels. Membrane replacement of this nature typically starts sooner than wholesale replacement. With more frequent and targeted replacements, plant performance based on an average membrane age can more smoothly be maintained and replacement costs spread out more than with wholesale replacement.

## 7.5 MAINTENANCE

Maintenance is of equal importance to the long-term successful operation of an SWRO plant. Maintenance considerations begin in the design phase when engineers include provisions for the regular maintenance of the plant and equipment. It continues through plant operation in preventing failures and prolonging system lifetime and performance.

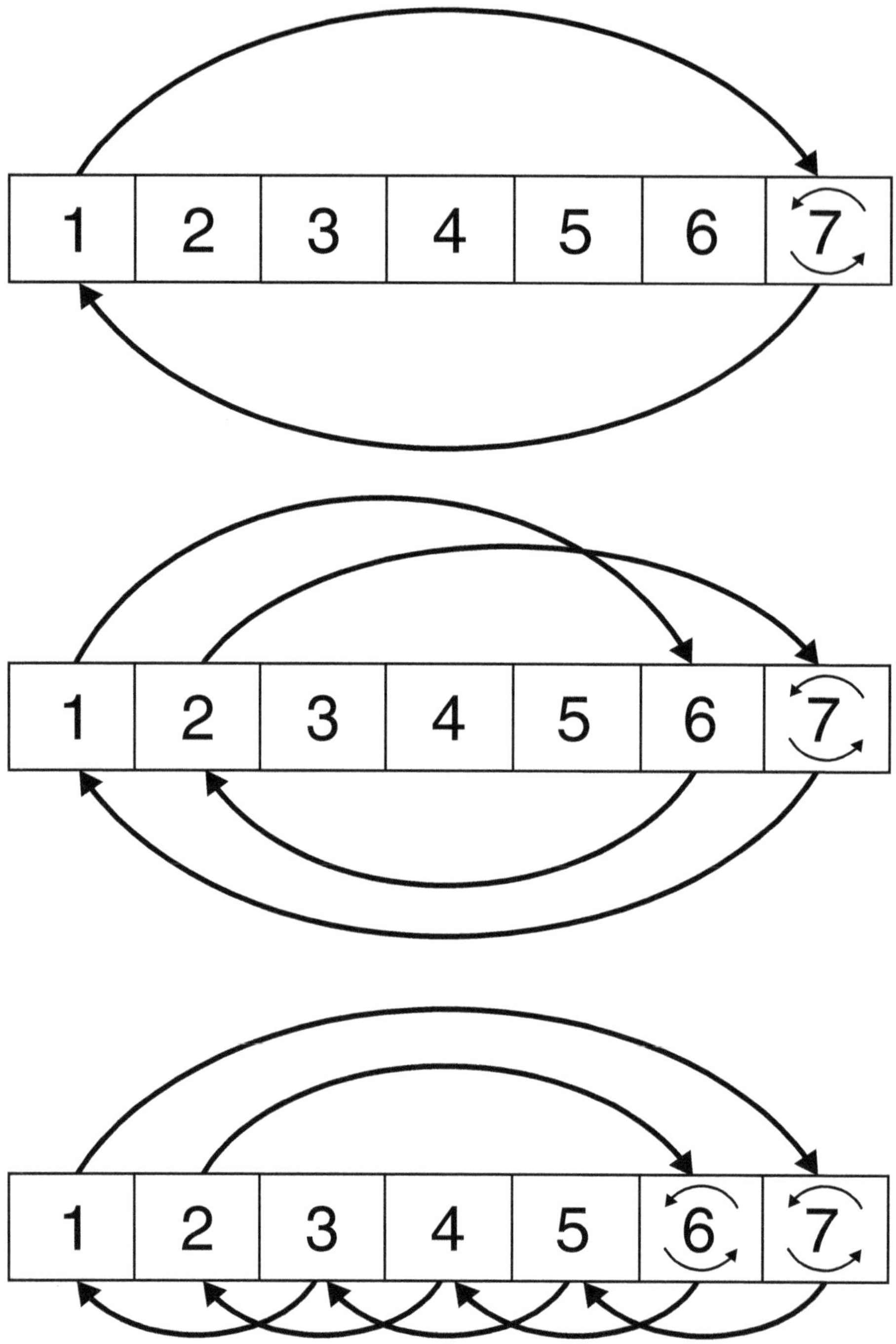


**Figure 7-4 Membrane relocation alternative method examples**

There are three major categories of maintenance: preventive maintenance, predictive maintenance, and corrective maintenance. Preventive maintenance is the routine maintenance of equipment that keeps it in operational capability (Figure 7-5). Predictive maintenance is the performance of ongoing parameter checks (such as vibration and temperature) that can be trended and/or have limits set beyond which item replacement is mandated to prevent unanticipated equipment failure. Corrective maintenance is any unscheduled task to troubleshoot, repair, or restore equipment that failed unexpectedly during operation.

*Source:* Iv3 Aqua Corporation

**Figure 7-5 Operators performing maintenance on an inline turbidimeter for raw water feed to a seawater reverse osmosis (SWRO) plant**

A successful maintenance program requires preventive and predictive maintenance items to be scheduled and performed continuously. This preventive and predictive maintenance is a scheduled organized plan set forth by the operators, engineers, or equipment manufacturers to prevent failures, preserve equipment life, and eliminate potential safety hazards.

The preventive and predictive maintenance plan depends on the type of equipment at the plant and reliability requirements. Table 7-5 summarizes typical routine maintenance activities for an SWRO plant and their frequency.

**Corrosion.** Seawater and SWRO concentrate are exceptionally corrosive fluids that can attack and corrode a wide variety of metals. Sunlight breaks down polymer coatings and even salt in the air aids in the corrosion of all SWRO plants. Corrosion control of structural steel members, tanks, frames, buildings, air-conditioning units, electrical conduits, and concrete is a critical daily maintenance item that should not be overlooked. Even the most high-tech plants with proper materials of construction can become unsafe or costly to maintain if corrosion is not continually addressed and prevented.

Most large SWRO plant maintenance is managed through a computerized maintenance management system (CMMS). The CMMS is a software suite that assists managers and maintenance personnel in planning, organizing, ordering, tracking, and documenting all maintenance items. CMMSs can range in simplicity from linking

**Table 7-5 Routine maintenance activities**

| Activity | Frequency |
|---|---|
| **General** | |
| Check for leaks | Daily |
| Wash down seawater leakage areas with freshwater | Daily |
| Corrosion prevention and control | Continuously |
| **Pumps** | |
| Clean strainers or remove flotsam from raw water tanks | As needed |
| Vibration, flow, and pressure checks | Daily |
| Check pump oil level | Weekly |
| Change pump oil and motor grease | Quarterly |
| Tighten mounting and piping bolts and check vibrations | Quarterly |
| Replace mechanical seals | Yearly |
| Inspect pump impeller | Every 2–5 years |
| **Electrical** | |
| Check motor current and temperature | Daily |
| Check transformer temperature and oil levels | Weekly |
| Inspect electrical gear for hotspots | Yearly |
| Tighten all electrical lugs | Yearly |
| **Valves** | |
| Exercise valves | At least quarterly |
| Inspect for leakage | As needed |
| Inspect valve, check valve seat | Yearly |
| **Tanks** | |
| Inspect tank internals and bottom for sedimentation | Yearly for chemical tanks, every 3–5 years for potable storage tanks |

equipment to spare parts inventory to automated parts ordering with neural network–aided predictive maintenance. CMMSs can generate work orders (WOs) organized by equipment and plant operational status, ensuring that a shutdown punch list of WOs is readily available to minimize downtime. Various KPIs are collected in each work order for upper management to review equipment and personnel trends. They can also provide tools for upper management to streamline maintenance across sites and perform root cause analysis on failures. Some CMMS programs can interface with existing emergency response plan (ERP) and SCADA programs, as well as provide web-based portals and cloud storage for system access by engineering and maintenance staff worldwide. The size and scope of a plant's CMMS should be based on staff needs, plant size, and process complexity.

## 7.6 CHEMICAL INVENTORY MANAGEMENT

Chemical inventory management is a crucial aspect of operations for a seawater desalination facility. This important task includes management of chemical purchases, storage, use, and disposal. A sound management plan will reduce environmental, health, and safety risks; maintain compliance with applicable laws; and provide increased safety, cost savings, and waste reduction that will improve the plant's overall performance.

As a rule of thumb, inventories of chemicals should be minimized whenever possible while maintaining the required redundancy for storage volumes and operational constraints. This reduces the potential for hazardous chemical spills, injury to personnel, or damage to property and lessens the amount of chemical disposal. The quantity of chemicals that must be kept on hand is determined by daily demand and lead times for new deliveries. Sufficient inventory should be maintained to cover the restocking interval. Generally, it is a good practice to maintain a supply of chemicals at the plant that is the full quantity of a cost-effective inventory replenishment so that when a replenishment arrives, there is a total stock of double the inventory. While 30-day supplies have traditionally been the norm in the United States, many desalination plants maintain supplies designed to last 2–4 months, depending on the plant location and the location from which the chemicals are being sourced.

The US Occupational Safety and Health Administration (OSHA) Hazard Communication Standard requires employers to inform workers about the identities and hazards of chemicals in the workplace. Federal regulations for chemical inventory management include

- providing a written plan and training for employees to handle chemicals appropriately,
- making SDS readily available for the chemicals being used, and
- conducting an inventory of hazardous chemicals.

A hazardous chemical is any that can cause a physical or health hazard. In a desalination plant, this could be oxidizing agents (e.g., sodium hypochlorite), corrosive substances (i.e., acids, bases), pressurized gases (e.g., carbon dioxide), or flammable liquids (e.g., polyvinyl chloride solvent cement). Hazards can be prevented by practicing good housekeeping and taking these precautions when storing chemicals.

- Keep storage, work areas, and walkways free of clutter and pests.
- Avoid stacking chemical containers to reduce the probability of spills.
- Keep chemical containers at least 6 ft away from hoist ways or floor openings and at least 10 ft away from exterior walls.
- Segregate chemicals that are incompatible.

- Provide adequate protection from the environment (e.g., storing powders indoors), noting that the storage of chemicals may take up as much as about 15 percent of the plot space of the plant.
- Provide adequate secondary containment in case a spill occurs.

Chemical inventory management software can help to track chemicals and maintain appurtenant records from receipt to disposal. These programs are widely available from many vendors and can be changed or customized to meet the specific needs of the facility. Key features of these products include system user access control, management of chemical purchasing and stocking information, and report-generation functionality.

## 7.7 SPARE PARTS MANAGEMENT

There are many executional options and metric-based approaches to spare parts management systems. More importantly, desalination plants are often required to have a high online availability at or near to full contracted/designed production, including installing redundancy so that item replacements can be installed "on-the-run." All spare parts management systems seek to balance the cost of maintaining spare parts inventory and the availability of those parts to maintain plant availability. Spare parts should be balanced between their value, shelf life, criticality to plant availability, and risk of failure. Having a CMMS or other system that allows part usage and failure rates to be tracked as well as a streamlined supply chain is beneficial to ensure proper spare parts management. An important consideration is the storage of spare parts, many of which will need to be in a secure, sheltered, and for some, air-conditioned environment. Storing spare parts will typically require a plot space of about 5 percent of the plot size of the whole facility.

## 7.8 PERSONNEL TRAINING

Staff training and education are important to allow plant personnel to gain a more complete understanding of the equipment they are responsible for operating and maintaining. Robust training programs are even more important when personnel do not already have experience with desalination plants or membrane treatment technology.

Training should include all aspects of the facility, including SOPs, safety, principles of operation, specific equipment maintenance, troubleshooting, and water sampling and testing. Besides company-generated training, most major equipment manufacturers also provide training for their equipment.

Ideally, personnel training should begin with the O&M staff present and be an integral part of the plant's startup and commissioning. This allows the staff ultimately responsible for the long-term success of the plant to shadow engineers, contractors, and specialists during this project phase, affording them a greater sense of ownership of the plant.

## 7.9 SAFETY

Access to sufficient quantities of safe water for drinking and domestic uses and for commercial and industrial applications is critical to health and well-being and the opportunity to achieve human and economic development. Desalination plants treat water impaired with salts (seawater and brackish water) or other contaminants as their sources. It is imperative that these plants operate safely regarding the people who ingest their water, plant operators, the environment, and the processing equipment.

Below is a safety listing of the various items, components, materials, and features encountered in a desalination plant. Membrane and thermal facilities have similar listings.

The fundamental differences are the replacement of the membranes, pressure vessels, and pretreatment sections of membrane plants with those of the vaporization chamber, recirculation loops, and heat exchangers for thermal desalination.

## 7.9.1 System Component Design

System equipment components should be designed to have a useful life (15–30 years) to the extent possible and practical. All plant components (equipment and structures) should be designed such that failures are repairable. Suppliers should provide documentation on individual equipment components that verifies the components are suitable for the water quality stated in the specifications.

Expected useful life should be detailed in the system specifications for the various mechanical and electrical equipment and devices. The system supplier, fabricator, and individual component manufacturer should provide documentation that these components meet these requirements. All system design and verification documentation should be summarized by the system supplier in the O&M manual.

## 7.9.2 Products and Components

**Materials of construction.** The structures and equipment components of a desalination system should be constructed of materials that are corrosion resistant in the environment in which they are placed. High-quality stainless steel and nonmetallic materials may be acceptable materials of construction, provided they can withstand the rigors of their internal and external environments. Generally, specialty stainless steel alloys as duplex and super-duplex alloys are chosen with consideration for high pressure versus cost and their benefits for a longer life cycle against corrosion/erosion and fouling. Industry procurement lead times and overall plant upgradability for future modifications or expansion should also be considered.

**Pressure ratings.** Many products and components may be used in systems with pressures that are other than ambient. In such instances, the materials of construction must meet the codes established by federal, state, and local regulatory agencies and professional associations like the American Society of Mechanical Engineers.

**Temperature variations.** Thermal systems and sometimes membranes operate under temperature conditions other than ambient. All materials in such use must be structurally unaffected by these temperature variations at maximum operating pressures.

**Toxicity levels.** All products and components must not be toxic to the public or the environment in their intended use. Where the water is for potable use, systems should be certified under NSF International Standards 60 and 61 and other standards promulgated by NSF International, AWWA, and ASTM International. The products and components must be inert to the environment when discarded according to regulatory standards.

**Other considerations.** All products and components must be safe to use. Protection of personnel, equipment, and the environment must always be assured through the proper monitoring and use with installation of relief valves and indicating, recording, and deactivation instrumentation.

**General plant safety components.**

- Emergency or area stop buttons
- Safety valves
- Guards, shields, and secondary containment
- Electrical breakers, fuses, and shunts
- SCADA
- Anchor bolts and pressure vessel straps for restricting movement

- Identification labels and numbering on equipment
- Color identification of electrical and flow piping with directional markings if needed
- Instrumentation test equipment
- Pressure vessel probing equipment

**Readily accessible plant procedural manuals.**

- Operating instructions
- Maintenance instructions
- Equipment specifications
- Raw material specifications
- Cost records for control
- Appropriate shift logbooks
- Safety manuals
- Emergency evacuation procedures
- Up-to-date equipment, piping, and instrumentation diagrams

**Safety stations.**

- Showers
- Spill-control stations
- Eyewash stands
- Containment dunes for flooding potential
- Stocked first aid stations

**Safety equipment.**

- Safety shoes and eyeglasses
- Hard hats
- Masks and respirators
- Clothing for handling hazardous chemicals and thermal conditions

### 7.9.3 Project Execution

The various components should be installed per the specification and contract. The system supplier should have a competent field representative on site at least for the following milestones:

- initial delivery of major equipment,
- beginning of the installation of the equipment (particularly membrane elements),
- final connections and flushing of all equipment, and
- other times as specified.

Membranes should be installed only after a thorough flushing of all piping to remove construction debris. All flushing water should be clean potable water and not allowed to sit in piping for extended periods. Wetted equipment not in daily operations should have provisions for a permeate flush to avoid stagnation of water with high levels of total dissolved solids.

Startup and commissioning should be the responsibility of the party responsible for overall project implementation and often with assistance from original equipment manufacturers for major components like membranes, pumps, and prefilters. The utility owner and the equipment suppliers should determine in advance which party will be responsible for operation of the system.

Startup should include electrical and mechanical checks of all equipment, leak detection, flushing, placing units in service, reconfirming the functional aspects of the system, and flow and performance verification. For a membrane plant, startup should only occur after the quality of the feed water to the SWRO system has met specifications—i.e., the pretreatment step has been activated and is functioning properly.

The system supplier should provide training to the O&M staff, which should consist of both classroom and hands-on activities. Training is a continuing activity. Existing staff should have scheduled periodic reviews and updates of all O&M procedures. These procedures should be written and readily available to personnel at all times. It is important that new O&M employees have training before they are certified to work on process equipment, instrumentation, and/or electrical supply.

Field-testing should include mechanical and electrical testing, which at a minimum should encompass equipment operation (pumps, valves, etc.), instrument verification, leak testing, power and signal connection verification, and programmable logic controller/personal computer program verification. A functional test of all operating equipment should be performed to verify that the equipment is performing safely.

## 7.10 RESILIENCY

A seawater desalination plant, like any other water treatment facility, is a critical asset of a water utility. Its function is to provide a reliable source of safe drinking water to the public. Steps should be taken to prevent risks (natural hazards and malevolent acts) from compromising the facility's infrastructure or contaminating its product water. This can be done by anticipating undesirable events or potential threats and having a plan in place to mitigate or prevent them, thereby increasing the resiliency of the desalination plant.

In October 2018, America's Water Infrastructure Act (AWIA) was signed into law and requires that community drinking water facilities serving more than 3,300 people conduct a risk and resilience assessment (RRA) and to develop an ERP within 6 months following an RRA. The new law requires that each RRA consider natural hazards and malevolent acts, resilience of water facility infrastructure, monitoring practices, financial systems, and chemical storage and handling. Based on the results of the assessment, each respective ERP must include strategies and resources to improve resilience, physical security and cybersecurity, response procedures for potential hazards affecting the safety of drinking water, implementation of measures to mitigate potential hazards, and strategies to detect natural hazards or malevolent acts that threaten the water system (*America's Water Infrastructure Act of 2018*).

In 2024, AWWA released its update for Standard G430, Security Practices for Operation and Maintenance. This standard provides important guidance to utilities in implementing security measures. Its purpose is to "define the minimum requirements for a protective security program for a water, wastewater, or reuse utility that will promote the protection of employee safety, public health, public safety, and public confidence" (AWWA 2024).

The following is a brief summary of AWWA G430-24, including a few of the many components relevant to properly securing a desalination facility. Readers of this manual are encouraged to consult the standard directly for more detailed security guidance. AWWA's G440-22 (Emergency Preparedness Practices) and OSHA's 29 CFR 1910.38 (Emergency Action Plan Standard) and 1910.120(q) (Emergency Response to Hazardous Substance Releases) are additional resources recommended for developing AWIA-compliant ERPs.

## Commitment to Security and Security Culture

An effective security program begins with a visible commitment to security by senior leadership of the utility. Leaders should establish a security culture by promoting security awareness throughout the organization. This may be accomplished by providing opportunities for employee suggestions, implementing training programs, issuing identification badges, incorporating security into job descriptions, measuring progress, and rewarding employees for positive behavior in enhancing security awareness. Specific roles and responsibilities should be assigned for creating, maintaining, and implementing the security plan. This plan should be explicit and easily communicated throughout the utility. Security should have a sustained focus within the utility by devoting appropriate resources to the topic in terms of budget and staff time.

## Assessment of Risk

A risk and resilience assessment should be conducted for all seawater desalination plants. In the case of a proposed facility, the assessment should be done proactively so that the findings of the assessment can be incorporated into the design and construction. The assessment should include, at a minimum, an asset characterization, threat characterization, consequence analysis, vulnerability analysis, threat analysis, risk and resilience analysis, and lastly risk and resilience management (for more details, see AWWA J100). The resultant assessment should be reviewed and updated periodically; the assessment should also be reviewed after significant events like new construction, receiving new information about a potential threat, or having sustained an attack or breach of security. The assessment should be integrated with other operational plans, including ERPs and business continuity plans, so that its priorities are considered and addressed in a broad context. These plans should be tested regularly through training, drills, or simulations.

## Access Control and Intrusion Detection

The security plan will require implementing various means of access control and intrusion detection to protect critical assets, including physical structures as well as information technology (IT) and SCADA systems. Areas for access control and detection should be identified with the risk and resilience assessment. Control and detection measures will be accomplished through installation of physical components and systems and through implementing various policies and procedures. This can include varying physical measures, like hardening walls and incorporating intrusion prevention devices on windows and doors; erecting fences and other barriers (Figure 7-6); providing locking devices for doors, gates, and hatches; and installing monitoring systems with intrusion alarms. The latter category may use motion detectors, enhanced lighting, or closed-circuit TV surveillance, among others. Surveillance methods should be implemented to detect chemical, biological, or radiological contamination through online monitoring systems and laboratory testing. This can be done directly or through surrogate monitoring. With IT and SCADA systems, physical protections may include installing and maintaining firewalls, separating business and operational systems, installing a system for virus protection, securing SCADA equipment locations, and incorporating encryption technologies.

## Policies and Procedures

Policies and procedures should be adopted to control access to critical assets by personnel through use of a hierarchical access card system or other means to maintain security. Staff should be required to display identification at all times, and visitor access should be controlled through a sign-in and escort system. A procedure for deliveries should include chain-of-custody control or tamper-evident packaging requirements. When

*Source:* Iv3 Aqua Corporation

**Figure 7-6 A reverse osmosis (SWRO) facility with security fencing and guard station**

legally permitted, background checks should be made on employees, and a protocol should be instituted to immediately rescind access privileges for employees who have been terminated or resigned.

### Communications

Strong internal and external communication strategies should be developed and implemented relative to security issues. Regular and ongoing communication with employees is critical to ensure that security matters are taken seriously, to promote employee safety during an event, and to enable effective employee participation when an event occurs. Effective communication with response organizations, regulatory agencies, and customers is also vital. Collaborative partnerships should be established with key agencies to ensure cooperation and effective coordination during emergency response and recovery. Mutual aid and assistance agreements should be reached with neighboring utilities and other relevant agencies.

## REFERENCES

American Water Works Association (AWWA). 2004. *Water Desalting Planning Guide for Water Utilities*. Denver: AWWA.

*America's Water Infrastructure Act of 2018*. 2018. 115th Cong., 2nd sess., *Congressional Record* 132: 3765–3893.

AWWA. 2007. Manual M46, *Reverse Osmosis and Nanofiltration*. Denver: AWWA.

ANSI/AWWA. 2024. Standard G430-24, Security Practices for Operation and Management. Denver: AWWA.

US Bureau of Reclamation. 2003. *Desalting Handbook for Planners*. Denver: USBR.

This page intentionally blank.

Chapter 8

# Case Studies

This chapter provides an overview of seven different desalination plants, including the three largest in the western hemisphere, three additional projects in the United States, and a facility in Israel that remains one of the largest in the world. Authors of this chapter made a concerted effort to maintain a standardized format across the case studies; however, due to the differences among the facilities, each case study communicates the most relevant and available information for its given location, resulting in a varied presentation structure.

# CASE STUDY: CLAUDE "BUD" LEWIS CARLSBAD DESALINATION PLANT (CDP)

**Management Approach:** Ocean discharge-colocation of desalination plant and power plant discharges

**Subcommittee Members:** Nikolay Voutchkov, Val S. Frenkel

**Project Location:** Carlsbad, Calif.

**Desalination Process:** Source seawater was originally supplied from the cooling water discharge of the Encina Power Station (EPS) and is now supplied from Agua Hedionda Lagoon (San Diego County Water Authority 2024). Pretreatment is by conventional single-stage, dual media gravity filtration (sand and anthracite) and source seawater conditioning by pH adjustment and antiscalant addition only when needed. Reverse osmosis (RO) separation is by a first-pass, single-stage membrane desalination system with a small partial second pass to achieve needed boron concentrations. Product water permeate posttreatment uses calcite contact filtration, pH adjustment, addition of carbon dioxide ($CO_2$), and disinfection using chlorination by sodium hypochlorite residual chloramine disinfection with the addition of ammonia. The product water is also fluoridated.

## Water Treatment Plant (WTP) Information

- Rated capacity: 50 million gal per day (mgd, or 189 million L per day [ML/d])
- Maximum concentrate flow: 54 mgd (204 ML/d)
- Typical production: 48–54 mgd (181–204 ML/d)
- Typical concentrate flow: 48–54 mgd

## Abstract

The CDP is located at the EPS in the city of Carlsbad. The CDP currently uses the existing power plant outfall to mix the concentrate discharge with the cooling water discharge from the power plant. The owner of the EPS had notified Poseidon Water (Boston, Mass.) of its plans to shut down and start decommissioning the EPS in late 2018. Since the existing intake configuration and regulatory approvals for the CDP are predicated on operation of the power plant and its associated cooling water flows, the transition to a standalone operation of the desalination plant will require permit modifications and upgrades to the EPS intake and discharge system.

## Process Design and Configuration

The CDP does not have separate intake and discharge. Instead, the desalination plant currently uses the existing cooling water discharge facilities of the EPS to collect seawater for freshwater production and to convey the desalination plant discharge to the ocean (see Figure 8-1). The desalination plant treatment incorporates single-stage granular media filtration followed by 20-µ cartridge filtration, first-pass seawater reverse osmosis (SWRO) with a partial second-pass membrane separation and posttreatment through calcite contact tanks and chlorination. The plant uses carbon dioxide and caustic soda for pH adjustment and ammonia addition for final conversion of the product water to residual chloramines. Fluoride is also added. The plant operates at 50-percent recovery, and energy recovery from plant concentrate is completed using pressure-exchangers (Energy Recovery, Inc., San Leandro, Calif.).

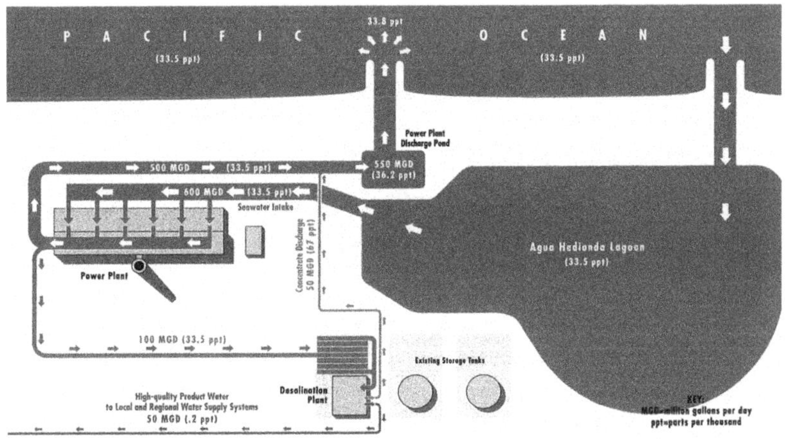


*Source:* Water Globe Consulting

**Figure 8-1 Schematic of the Claude "Bud" Lewis Carlsbad Desalination Plant system**

## Project Background

Historically, San Diego County relied on water imported from Colorado River and the bay delta for a majority of its drinking water supply. The 50-mgd (189 ML/d) CDP now provides approximately 8–10 percent of the total county water supply. The project provides a local drought-proof water supply that is not dependent on snowpack or rainfall and reduces the need for water exports to San Diego County from Colorado River and the California State Water Project.

As shown in Figure 8-1, under typical operational conditions, a minimum of 304 mgd (1150 ML/d) of seawater enters the power plant intake facilities, and after screening, it is pumped through the plant's condensers to cool them and thereby remove the waste heat created during the electricity-generation process. The CDP intake structure connects to this discharge canal and under normal operational conditions, diverts 100 mgd (378 ML/d) of the 600 mgd (2271 ML/d) of cooling water for production of potable drinking water. Approximately 50 mgd (189 ML/d) of the diverted cooling seawater is converted to fresh drinking water via RO membrane separation. The remaining 50 mgd has salinity approximately two times higher than that of the ocean water (67 parts per thousand versus 33.5 ppt). This seawater concentrate is mixed with approximately 4 mgd (15 ML/d) of backwash water from the dual media filters (DMFs). The combined discharge with a salinity of approximately 65 ppt returns to the power plant discharge canal downstream of the point of intake for blending with the remaining cooling water before conveyance to the Pacific Ocean. Under average conditions, the blend of 200 mgd (757 ML/d) of seawater and 54 mgd (204 ML/d) of concentrate and backwash has a discharge salinity of 40 ppt. The blend will be discharged to the Pacific Ocean through the existing power plant ocean discharge structure, which extends approximately 700 ft (213 m) from the shore (see Figure 8-1).

## Permitting and Regulatory Overview and Procedure

The environmental review process for this project included four key components: Environmental Impact Report certified by the City of Carlsbad, National Pollutant Discharge Elimination System (NPDES) discharge permit approved by the San Diego Regional Water Quality Control Board, coastal development permit issued by the California Coastal Commission, and the site lease for plant intake and outfall provided by the California State Lands Commission.

Construction and operation of the CDP required obtaining permits or approvals from a number of regulatory agencies:

- California Environmental Quality Act (CEQA) (Note: the City of Carlsbad was the CEQA Lead Agency.)
- NPDES permit from San Diego Regional Water quality control board
- Intake and discharge lease from California State Lands Commission
- Coastal development permit from the California Coastal Commission
- Consultation with the US Fish and Wildlife Service and National Oceanic and Atmospheric Administration Fisheries in accordance with Section 7 of the federal Endangered Species Act
- Consultation with California Department of Fish and Wildlife through California Fish and Game Code Section 2081 for state-listed threatened or endangered species
- Incidental Harassment Authorization under the Marine Mammal Protection Act for potential impingement of marine mammals on the screens of the power plant in case the power plant discontinues the use of once-through cooling for its generation facilities and the desalination plant becomes the sole user of the existing power plant intake

## Analyses of Plant Concentrate

CDP concentrate has a salinity of 60–67 ppt. The dilution of the combined concentrate/effluent discharge was modeled using a tridimensional hydrodynamic model. This CDF model used for these projects was initiated for four sets of average and extreme environmental conditions: a worst-case day, an average day, a worst-case month, and an average month. For each of these conditions, the model analysis studied dispersion and dilution of the concentrate for a 54-mgd (204 ML/d) desalinated water production. The key variables controlling the concentration and persistence of the elevated salinities power plant–desalination plant discharge used in these models are desalination plant and flow rates, ocean salinity, ocean temperature, ocean water levels, mixing wave action, mixing current action, and mixing wind action.

Aquatic life salinity tolerance threshold was determined as a part of the concentrate environmental studies. All marine organisms are naturally adapted to changes in seawater salinity. These changes occur seasonally and are mostly driven by the evaporation rate through the ocean surface, by rain or snow events, and by surface water discharges. The natural range of seawater salinity fluctuations could be determined based on information from sampling stations located in the vicinity of the discharge and operated by federal, state, or local agencies and research centers responsible for ocean water quality monitoring. Typically, the range of natural salinity fluctuation is at least +/– 10 percent of the average annual ambient seawater salinity concentration. For example, based on ocean water quality monitoring data collected for the development of environmental impact reports for the CDP, the average annual ocean water salinity (i.e., total dissolved solids [TDS]) concentration is 33,500 mg/L, while the maximum salinity reaches 36,800 mg/L, which is approximately 10 percent above the average. The "10 percent increment above ambient ocean salinity" threshold is a conservative measure of aquatic life tolerance to elevated salinity. The actual salinity tolerance of most marine organisms is significantly higher than this level. Detailed analysis and salinity tolerance studies performed for the CDP indicate that most species inhabiting the Southern California Bight and especially those found in the discharge areas of the two plants have a salinity tolerance threshold of 40,000 mg/L or higher. (The Southern California Bight is an open embayment extending from Point Conception, Calif., into Baja California, Mexico, 125 mi [201 km] offshore.)

### Economic Evaluation

A new mile-long (1.6-km long) outfall pipeline was considered as an alternative to the use of the existing open power plant outfall structures that extend approximately 700 ft (213 m) from the shoreline. Construction of new outfall structure was not found to be viable because of the potential negative environmental impact on existing kelp beds. Concentrate discharge using subsurface wells was not found feasible due to the unfavorable hydrogeological conditions in the vicinity of the plant and the significant environmental impacts of construction and maintenance of numerous discharge wells along the shore.

### Next Steps for Standalone Operation

In addition to the intake and discharge system modifications that were required to accommodate the transition to standalone operation, additional intake and discharge system improvements were required to comply with the California Ocean Plan Amendment adopted by the state Water Resources Control Board in May 2015. This new regulation establishes requirements for the intake of seawater and discharge of brine from desalination facilities in California.

Lastly, the standalone permit modifications and intake and discharge improvements and related permits have been structured to provide for a potential 10-percent increase in CDP production capacity available due to improved RO membrane technology.

The CDP modifications consist of changes to the project as constructed and operating. The changes involve installation of new intake screens in Agua Hedionda Lagoon, a fish-friendly pumping structure, auxiliary facilities, and minimal improvements to the existing intake and discharge tunnels.

The CDP process and dilution water passes through new wedgewire screens located in Agua Hedionda Lagoon. The screens are designed to minimize impingement mortality and reduce entrainment mortality through the use of 1-mm slot openings and a through-screen velocity of less than 0.5 ft/s (0.15 m/s). After screening, the existing intake pump station continues to deliver the process water to the CDP for processing through the pretreatment and RO membrane desalination system. Approximately half the water volume processed by the CDP leaves the CDP as potable drinking water and the other half as concentrated seawater with approximately twice the original intake water salinity.

A portion of the daily 299 million gal (1,132 ML/d) of water passing through the intake screens is transferred to the discharge tunnel by the low-impact fish-friendly pumps to reduce the salinity of the brine discharge. The combined flow is discharged to the ocean for further dilution in the brine mixing zone.

The EPS was decommissioned in 2018. The transition to standalone operation was completed in three phases: continued use of the EPS screens and pumps in 2019, operation of the EPS screens in conjunction with the fish-friendly dilution pumps in 2020–2022, and operation of the permanent intake and discharge modifications commenced in 2023.

## REFERENCE

San Diego County Water Authority. 2024. *Seawater Desalination*. www.sdcwa.org/your-water/local-water-supplies/seawater-desalination/ (accessed June 20, 2025).

# CASE STUDY: ESCONDIDA WATER SUPPLY (EWS) PROJECT

**Management Approach:** Deep ocean intake/discharge, greenfield project

**Subcommittee Members:** Greg Wetterau, Christian Sanders

**Project Location:** Antofagasta, Chile

**Desalination Process:** Raw water, sourced from the deep (approximately 85 ft, or 26 m) open ocean intake, is dosed with acid, ferric chloride, and polymer (as an option) and then filtered through two-stage dual media pressure filters following in-pipe hydraulic flocculation. As standard, cartridge filters are installed between the DMFs and RO trains to protect the membranes. Single-pass RO equipped with dual work exchanger energy recovery (DWEER) devices achieves the energy consumption and treated water quality objectives for this project. Posttreatment includes the addition of corrosion inhibitor (zinc orthophosphate), carbon dioxide, and soda ash dosing. Sodium hypochlorite is dosed for disinfection.

## WTP Information

- Rated capacity:
  - Initial (2017) = 2,500 L/s (approximately 55 mgd)
  - Upgraded (2019) = 3,200 L/s (approximately 73 mgd)
- Maximum concentrate flow: 7,033 L/s (approximately 160 mgd)
- Typical production: 2,500–3,200 L/s (55–73 mgd)
- Typical concentrate flow: 6,271–7,033 L/s (143–160 mgd)
- Pretreatment filtration rate:
  - 1st Stage = 9.3 m/h (all filters online)
  - 2nd Stage = 13.8 m/h (all filters online)
- Design RO recovery = 50 percent
- Design RO flux = 13.2 L per square meter per hour (LMH) (7.8 gal per square foot per day [gfd])

## Abstract

The EWS project, owned and operated by BHP, is the largest desalination plant in the Americas. The plant provides water for BHP's mining operations at the Escondida copper mine located in Chile's Atacama Desert. In addition to the desalination facility itself, the project also includes four high-pressure pump stations and over 100 mi (161 km) of two 42-in. pipelines that convey the treated water to the mine site located approximately 3,100 m (approximately 10,200 ft) above mean sea level.

## Project Background

Due to years of overextraction, most major mining operations in northern Chile are faced with significant future reductions to local groundwater supplies, which have historically been the primary source of water used in the region. In an effort to secure and accommodate future expansion of the Escondida copper mine, the world's largest copper producer, BHP turned to desalination to fully meet its current and future water needs for the facility by 2030.

Notable achievements of this project included overcoming significant site layout constraints for this large-scale facility and the construction of a complex conveyance system that moves the treated water to an elevation slightly greater than 10,000 ft (3,048 m) above sea level over a distance of more than 100 mi (roughly 160 km) in one of the most seismically active regions in the world.

## Description of the Proposed Solution

The EWS facility contains four RO plants that each produce around 800 L/s (approximately 18 mgd) while operating at 50-percent recovery. A hybrid RO array operating as a single pass is sufficient to meet the permeate salt rejection requirements for the project. Raw water is sourced from a deep open ocean intake to reduce the pretreatment requirements and minimize the risk to production from algal bloom events, which are known to occur in the area. Two-stage dual media pressure filters remove organic and colloidal material bound in floc created by coagulant dosing (with ferric chloride). Under normal operating conditions, the second stage filters would only be expected to provide a minor polishing step; however, during poor raw water events, with shorter filter run times and more frequent backwashing, filter particle breakthrough from the first stage filters and/or increased frequency of filter ripening would be expected to place more load on the second stage.

Because of the critical role this facility plays in copper production at the mine, the design incorporates redundancy features, notably a redundant intake shaft that allows one tunnel to be fully offline for maintenance without affecting plant production. Shock chlorination at the intake towers using sodium hypochlorite is used to control biofouling in the shafts as well as raw water and pretreatment infrastructure. Sodium bisulfite can be dosed pre- or post-cartridge filters to quench any remaining free chlorine residual before the RO membranes.

In the area of the Chilean coast where the plant is located, land generally rises vertically from the ocean, providing a very restricted surface area for large-scale construction. To reduce the plant footprint, the design utilizes pressure filters that are close-coupled with the RO to eliminate the need for intermediate pumping and storage between the two process units. Close coupling provides the additional advantages of energy savings and elimination of a potential establishment site for biofouling. For backwashing, the pressure filters utilize RO brine to eliminate filtered water consumption for cleaning and thus reduce the overall surface area required for filtration.

RO trains, along with the energy recovery devices (ERDs), are vertically arranged such that temporary scaffolding or mobile elevated work platforms are required to access many of the pressure vessels, which is common on many large-scale plants.

As the desalinated water is not intended for potable use, posttreatment of the RO permeate is only necessary to minimize corrosion potential and prevent biofilm growth in the conveyance infrastructure. Chemical dosing includes zinc orthophosphate as a corrosion inhibitor, carbon dioxide and soda ash for pH and alkalinity adjustment, and sodium hypochlorite for biofilm control.

## Permitting and Regulatory Overview

Due to the sensitivity of the nearby marine environment and the greenfield nature of this large-scale desalination project, permitting for the intake/outfall facilities required the implementation of several novel approaches to the offshore and underground construction of the infrastructure. For example, advanced slurry tunneling machines were used to bore directly from the shore out to where the vertical shafts associated with the intake and outfall structures would be located. To construct the vertical shafts through a seabed comprised of hard rock, the largest diameter offshore drill in the world had to be employed.

The intake tower design minimizes entrainment of aquatic life by maintaining a maximum screen velocity of <0.1 m/s (0.3 ft/s) at the maximum instantaneous raw water flow rate. In addition, a 150-mm spacing between bars is used to exclude the passage of debris and large fish through the screens.

# CASE STUDY: POINT LISAS SEAWATER DESALINATION FACILITY

**Management Approach:** Discharge of concentrate to oceans

**Subcommittee Members:** Greg Wetterau, John Thompson

**Project Location:** Point Lisas Industrial Estate, Trinidad and Tobago

**Desalination Process**: Supplied from open shallow intake in Gulf of Paria. Pretreatment is a combination of conventional single-stage, dual media, gravity filtration (sand and anthracite) and microfiltration (MF) using outside-in polyvinylidene difluoride (PVDF) membranes. RO includes a two-stage first pass followed by a two-stage approximately 90-percent second pass. Product water permeate posttreatment uses lime and carbon dioxide; disinfection uses sodium hypochlorite.

## WTP Information

- Original capacity: 28.8 mgd (109 ML/d), concentrate flow: 27.8 mgd (105 ML/d)
- Maximum production: 54 mgd (204 ML/d), concentrate flow: 52 mgd (196 ML/d)
- Typical production: 48 mgd (182 ML/d), concentrate flow: 46 mgd (174 ML/d)

## Background

The Point Lisas seawater desalination facility in Trinidad, West Indies, was first commissioned in 2002 with a capacity of 28.8 mgd (109 ML/d). The plant has undergone seven separate expansions and now has a rated capacity of 48 mgd (128 ML/d), operating near capacity throughout the year. The water produced is supplied to local industrial customers and for broader domestic supply by the Trinidad and Tobago Water and Sewerage Authority (WASA), maintaining a maximum 85 mg/L dissolved solids. Industrial uses, which constitute 30–50 percent of the plant production, are primarily for process water and boiler feed for ammonia, methanol, and iron and steel plants in the Point Lisas Industrial Estate. The plant provides drinking water to 25 percent of Trinidad's population, with WASA purchasing water at an agreed rate of about US\$1/$m^3$ in 2021.

The desalination facility is owned and operated by the Desalination Company of Trinidad and Tobago (Desalcott), a single purpose entity formed in 1999 to sell desalinated seawater to WASA. Desalcott was a joint venture enterprise between Trinidadian company Hafeez Karamath Engineering Services (HKESL) with a 60-percent shareholding and GE Water/Ionics (now part of Suez) with a 40-percent shareholding. In November 2012, HKESL purchased the entire GE Water shareholding, making Desalcott fully locally owned and operated. The plant maintains an operations, maintenance, and management staff of 120 local employees.

## Water Source

Raw seawater from the Gulf of Paria is screened and pumped 1.5 km (0.9 mi) to the plant site via two 42-inch–diameter pipelines. Seawater composition in this region is strongly influenced by runoff from the Orinoco River delta about 10 mi (16 km) away in Venezuela. Challenges presented are widely variable turbidity and salinity during the rainy season and high salinity during the dry season. Salinity averaged 32,400 mg/L TDS in 2020 with a maximum of 37,900 mg/L during early summer. Turbidity averaged 3.6 ntu, with a peak of 22 ntu during the month of December. Historically, turbidities as high as 70 ntu have been recorded. Water temperature ranged from 28 to 32°C (82–89°F) throughout the year, without a large degree of seasonal variation.

With climate changes and reduced rainfall patterns, there has been a marked change in salinity and temperature of the raw water. Abnormal hurricane seasons, increased

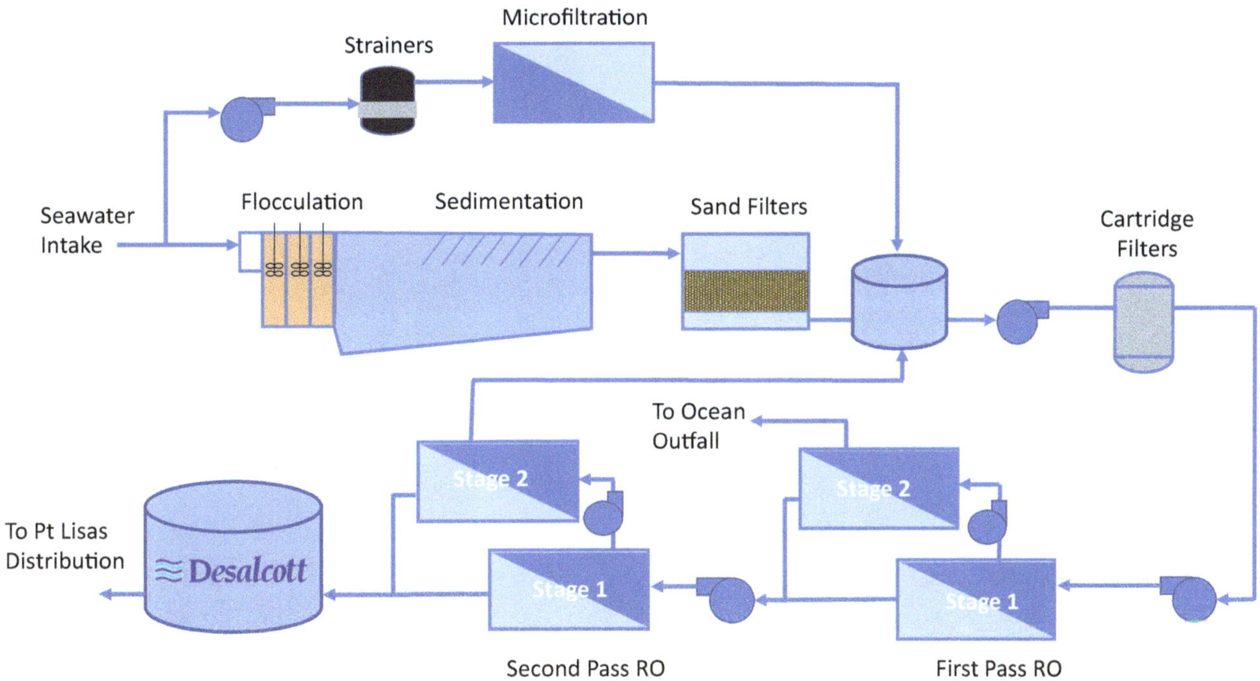


*Source:* Desalcott

**Figure 8-2 Process flow diagram for Point Lisas Desalination Plant**

temperature, and changing flows from the Orinoco River and local rivers in Trinidad are also believed to be changing the nature of foulants in the water.

## Pretreatment

The pretreatment includes two separate process trains, the original employing ferric chloride coagulation with media filters, and the other employing inline strainers and MF. A process flow diagram showing both process trains is included in Figure 8-2.

The conventional filtration process handles around 90 percent of the influent flow and includes a rapid mix tank, four flocculation/sedimentation basins, and 10 dual media sand/anthracite filters. The sludge is removed and further processed in a thickener for eventual disposal as a cake. The filters are backwashed with the brine from the SWRO plant.

The MF process handles around 10 percent of the influent flow and includes 150-µ strainers (Arkal Filtration Systems, Bet Zera, Hazafon, Israel) and five independent racks of PVDF membranes, supplied by Pall Corporation (Port Washington, N.Y.). The flows from the conventional filters and MF are combined in a filtered water clearwell before being pumped through cartridge filters and on to the SWRO systems. A photo of the MF trains is included as Figure 8-3.

Shock chlorination sequences using sodium hypochlorite are used intermittently to minimize/prevent biofouling of the RO membranes from the biological activity in the seawater. Sulfuric acid is utilized to prevent carbonate scaling.

## Desalination

The desalination plant includes eight seawater RO trains followed by eight brackish water RO (BWRO) trains to provide high purity water. Two-stage centrifugal pumps directly

*Source:* Desalcott

**Figure 8-3 Point Lisas microfiltration (MF) system**

connected to an electrical motor provide pressurized seawater for the SWRO units. The motor is directly connected to a Pelton energy recovery turbine that is driven by the RO reject stream to recover energy and reduce the power required for pumping. The SWRO trains (Figure 8-4) are two-stage units operating at 53-percent recovery with an interstage booster pump to balance flow between stages. Permeate is conveyed to the pass-two clearwell before being pumped to the BWRO units.

BWRO trains are two-stage units operated at 94-percent recovery with interstage booster pumps to balance flow between stages. Second-pass permeate is conveyed to the product tank.

SWRO brine is conveyed to a waste outfall wet well, where a portion is used to backwash the DMFs and the remaining flow is conveyed to the ocean outfall. Second-pass RO concentrate is returned to the filtered water clearwell.

## Posttreatment

The final product is treated with sodium hypochlorite for disinfection and lime to provide hardness and a stable pH of 7–8.5.

*Source:* Desalcott

**Figure 8-4 Point Lisas seawater reverse osmosis (SWRO) system**

# CASE STUDY: SAND CITY COASTAL DESALINATION PROJECT

**Management Approach:** Beach well supply, subsurface concentrate discharge

**Subcommittee Member:** Greg Wetterau

**Project Location:** Sand City, Calif.

**Facility Operator:** California American Water

**Desalination Process**: Water is supplied from vertical beach wells. Pretreatment is limited to cartridge filters. Positive displacement feed pumps allow for constant production with variable feed water salinities. Pressure exchanger ERDs are used with single stage RO. Ultraviolet (UV) light disinfection, chlorination, and a calcite contactor provide posttreatment.

## WTP Information

- Installed capacity: 0.6 mgd (2.3 ML/d); concentrate flow: 1.8 mgd (6.8 ML/d)
- Typical production: 0.3 mgd (1.1 ML/d)
- Completion date: June 2009
- RO recovery: 33 percent
- Design flux: 8.9 gfd (8.9 LMH)

## Background

With a population of only a few hundred people, Sand City, Calif., has historically been a largely industrial area. The city has been redeveloping into a mixed-use community to attract growth and new residents. In order to make this redevelopment possible, Sand City needed a reliable water supply that was not subject to periodic reductions due to drought and environmental pressures. With all of the city's water supplied through California American Water, the Seaside Aquifer and Carmel River have become overdrafted.

Sand City initiated a 0.6-mgd (2.3 ML/d) design–build (DB) project to facilitate the treatment of seawater to domestic drinking water standards. The project included a fully operational water treatment facility and associated permits. CDM Smith, Inc., was hired to provide DB services for this project. The proposed project augmented previous water supplies as the city grew and redeveloped.

A key challenge of this project was securing the permits for the first full-scale seawater desalination plant in California permitted under the state's new surface water treatment regulations. To minimize challenges and ensure that permitting would be secured, a conservative approach was taken on the project. Efforts included the use of subsurface intakes rather than drawing from an open intake source, the use of a subsurface brine discharge well, limiting the salinity of brine to below seawater salinity, and providing filtration and disinfection for *Giardia, Cryptosporidium*, and viruses.

The plant draws its source water from beach wells, returning waste concentrate to subsurface discharge wells. The treatment facility includes UV disinfection plus RO to provide disinfection credits in keeping with the Long Term 2 Enhanced Surface Water Treatment Rule (LT2). As the first full-scale seawater desalination plant in California permitted under the LT2 surface water treatment regulations, the Sand City Desalination Facility (Figure 8-5) set precedent for integrity testing and performance requirements for future, large-scale projects, such as Carlsbad and Santa Barbara.

## Treatment Process

The plant utilizes RO technology to treat influent water, which is pumped from nearby shallow beach wells. After passing through the RO system, the treated water is run

*Source:* Greg Wetterau

**Figure 8-5 Sand City (Calif.) Desalination Facility**

through a UV system to gain compliance with the recently enacted LT2. Product water stabilization is provided through calcite contactors, caustic soda (sodium hydroxide), and blending with existing water from the California American Water water distribution system. The treatment process includes cartridge filters and RO for primary treatment (Figure 8-6); UV for disinfection and redundant destruction of any bacteria or protozoa present in the feed water; chlorination for redundant virus inactivation and secondary disinfection; and finally, carbon dioxide, calcite contactors, caustic soda, and blending with existing distribution system water for product water stabilization. Concentrate from the desalination process is blended in order to meet coastal commission permit requirements then sent to a subsurface discharge well.

## Energy Efficiency

By selecting positive displacement pumps operating with 95-percent efficiency, the facility is able to adapt to the highly variable source water salinity (ranging from 17 to 28 g/L) without the use of variable frequency drives, reducing capital costs, operating costs, and the energy footprint. The use of ERDs further reduces the overall pumping costs, resulting in a 67-percent decrease in the size of the RO feed pumps, which thus greatly reduces the carbon footprint of the facility. Power usage for the first year of plant operation (2010) averaged 5.2 kW•h per 1,000 gal (1.4 kW•h /$m^3$); however, this number increased to 8.5 kW•h per 1,000 gal (2.2 kW•h/$m^3$) by 2012 due to increasing salinity in the source water.

*Source:* Greg Wetterau

**Figure 8-6 Sand City, Calif., reverse osmosis (RO) system**

## Posttreatment

The posttreatment of the permeate was critical because of the variable water qualities already present in the drinking water distribution system, the limited footprint available for posttreatment, and the goal to have the facility remotely operated. Permeate is disinfected with UV light and sodium hypochlorite, injected with carbon dioxide, and then passed through calcite contactors to add calcium hardness back into the permeate stream. Water from the drinking water distribution system is then routed through the plant, and up to a 50-percent blend of this system water provides additional buffering to the water. A final pH trim with caustic soda assures that the stability and aggressiveness of the blended product are mitigated. By utilizing up to a 50-percent blend with existing system water, the blending helps assure that the product water from the desalination plant matches the existing, highly variable system water while it reduces the risk of corrosion within the distribution system piping and stabilizes chemical costs.

## Permitting

In order to reduce the permitting challenges, particularly regarding requirements of the California Coastal Commission and the Regional Water Quality Control Board, the Sand City project followed a conservative approach and minimized the risk of environmental and public impacts. Specifically, subsurface intakes or beach wells were used to avoid impingement and entrainment (I&E) of sea life associated with open intakes, and subsurface brine discharge wells were used, with the concentration of TDS in the brine limited to levels lower than the surrounding seawater. To simplify the monitoring requirements and California Department of Public Health (now the Division of Drinking Water) approval process, the Sand City facility was designed to provide the highest level of treatment required for an impaired water body, avoiding some of the requirements for an extensive source water monitoring program and a full watershed sanitary survey.

# CASE STUDY: CHARLES MEYER DESALINATION FACILITY

**Management Approach:** Discharge of concentrate to oceans

**Subcommittee Members:** Val S. Frenkel, Greg Wetterau

**Location:** Santa Barbara, Calf.

**Desalination Process:** Intake ocean intake water is pretreated with two sets of media filters, then pressurized for single-pass RO before being distributed to customers.

## WTP Information (Ferguson 2004)

- Original capacity: 6.7 mgd (25.4 ML/d), maximum concentrate flow: 12.5 mgd (47.3 ML/d)
- Current capacity: 2.8 mgd (10.6 ML/d), maximum concentrate flow: 3.4 mgd (12.9 ML/d)

NOTE: Current maximum concentrate flow calculated here is based on current capacity and on communication from the county stating that "approximately 45 percent" of the seawater treated at the plant becomes potable water (County of Santa Barbara Public Works 2024).

## Summary

An existing wastewater outfall was selected as an obvious disposal method that would provide initial dilution and minimize time and cost of construction. The plant was in long-term storage from 1992 to 2015. Regional partners had dropped out, and capacity had been sold off. However, the facility was retrofitted and restarted in 2017 at a reduced capacity. The facility is now fully owned by the City of Santa Barbara (Ferguson 2004).

## Process Design and Configuration

Ocean water is pumped at a very low pressure through a 2,500-ft (762-m) seawater intake line, an abandoned wastewater discharge line that was relined with high-density polyethylene pipe (Ferguson 2004). The incoming seawater is pretreated in sets of round horizontal media filters: primary filters consisting of sand, gravel, and anthracite, and secondary filters consisting of the same media as primary plus garnet. Cartridge filters complete the pretreatment.

Pretreated seawater is pressurized to 800 psi (5,500 kPa) for a single pass through RO membranes. Modular desalination units were installed in trailers. Each "unit" consists of a pumping trailer and a membrane trailer and is capable of producing 625 acre ft/year (203 million gal/year or 770,938 $m^3$/year) of desalinated water (Figure 8-8 and Ferguson 2004).

Finished water is pumped into an existing water main for distribution to water customers. The concentrate is combined with treated wastewater from the adjacent wastewater treatment plant (WWTP) and discharged to the ocean at the end of the 1.5-mile-long (8-km) outfall line (County of Santa Barbara Public Works 2024). The outfall line includes a diffuser.

It takes approximately 6,600 kW•h of electrical energy to produce 1 acre ft (326,000 gal or 1,234,000 L) of desalted water (County of Santa Barbara Public Works 2024).

## Initial Project Background

Multiple factors led to the construction of the Charles Meyer desalination facility, including extreme drought conditions in the late 1980s that pushed the region into a long-term supply crisis. In response, the city requested proposals for a new water supply, eventually leading to the approval for a seawater desalination facility. The city entered a design–build–operate contract with Iconics, Inc., and built the Charles Meyer facility under emergency conditions following this timeline:

- Request for proposal released in April 1990;
- Iconics, Inc., proposal selection in August 1990;
- Proposed facility plan's environmental review;
- Permits granted from Coastal Commission, Regional Water Quality Control Board, and US Army Corps of Engineers;
- Overlapping permitting, design, and construction;
- Notice to proceed with construction issued in May 1991; and
- Facility officially completed in March 1992.

The plant was then operated from March to June of 1992 before being taken offline due to abundant rainfall during the winter of 1991–1992.

While the facility had originally been planned as an emergency water supply, city voters elected to make it a permanent supply in 1991, maintaining the plant as a backup during future droughts. The facility also had the potential for use during nondrought periods to meet regional or statewide needs for water by operating under a water exchange agreement. To obtain permanent status, the facility went through additional environmental review and permitting, which was completed in October 1996, with the Coastal Development Permit issued by the California Coastal Commission for a permanent facility producing up to 10,000 acre ft per year (3,260 million gal/year or 12,330,000 m$^3$).

The facility remained dormant for nearly 20 years until an exceptional drought in 2014–2015 prompted the Santa Barbara City Council to reactivate the plant. In May 2017, after a US$72-million retrofit led by IDE Americas, the plant was restarted with a reduced capacity of 2.8 mgd (10.6 ML/d). The retrofit included a new screened intake structure with 1-mm wedge wire screens, diffusers for the brine discharge, and high-efficiency pumps and motors (see Figure 8-7) to reduce the plant electrical load by 40 percent (City of Santa Barbara 2024; Ferguson 2004).

## Permitting and Regulatory Overview and Procedure

Current permitted flow rates are outlined in the City of Santa Barbara's NPDES Permit No. CA0048143 for the El Estero Water Resources Center:

> The average monthly rate of combined discharge of brine and other supplemental desalination flows from the desalination facility shall not exceed 12.5 MGD when the desalination facility is operational. Supplemental desalination flows shall be limited to startup maintenance flows, off-spec treated seawater, treated filter backwash water including dewatering of backwash and lime clarifier solids, and potable water. These flows may be periodically discharged during initial facility start-up, during or after facility maintenance, or at other times when the desalination facility is operating or not otherwise delivering potable water. The term desalination brine throughout the Order includes these supplemental desalination flows. Temporarily discharging such supplemental desalination flows to the Pacific Ocean does not constitute a "bypass" as defined in Attachment D, Standard Provision I.G.1.a of this Order.

*Source:* City of Santa Barbara, Calif.

**Figure 8-7 Containerized treatment systems used in 2017 plant retrofit (City of Santa Barbara 2022)**

Disinfected and dechlorinated secondary treated wastewater and waste brine are discharged to the Pacific Ocean through an 8,720-foot outfall system with diffusers located along the terminal 720 feet of the outfall. The outfall terminates in the Santa Barbara Channel (34.39194° N latitude; 119.6675° longitude) in approximately 70 feet of water. The minimum initial dilution ratio of the outfall/diffuser system is 120 to 1 (seawater to effluent) when no brine discharge occurs. When brine discharge does occur, the minimum initial dilution ratio is 44 to 1 (seawater to effluent). The hydraulic capacity of the outfall is 28 MGD. (California Regional Water Quality Control Board, Central Coast Region 2019)

### Economic Evaluation

The city's facility was originally built by a private company, Ionics, Inc., under a "take or pay" contract. Over the 5-year contract period, the city, along with the Montecito and Goleta Water Districts, paid off the US$34-million construction cost and either paid for water produced or paid to maintain the facility in standby mode.

The cost of desalted water was approximately US$1,100 per $ft^2$ including labor, chemicals, power, maintenance, and a sinking fund to replace worn components.

The 2017 retrofit was built by IDE under a design–build–operate contract, with the capital cost financed over 20 years with a 1.6 percent low-interest loan, equating to US$4.2 million per year in debt service. Annual operating costs are estimated to be US$4.1 million at full production and US$1.5 million in nonoperation or standby mode.

### Key Project Lessons Learned

Due to the drought emergency, time was a critical factor. Getting power to the site (66kV) was also a significant obstacle (Bjork 2007).

Being the first large seawater desalination project to go through the California permitting process has its advantages, especially given the emergency motivation. Subsequent projects may face additional scrutiny and more stringent conditions.

Taste issues were reported by some customers. It is unclear if these issues were significant, given that some complaints came from customers who were not receiving desalinated water (Bjork 2007).

## ACKNOWLEDGEMENT

Malcolm McEwen developed original case study on the City of Santa Barbara, Calif., in 2008.

## REFERENCES

Bjork, R. 2007. Personal communication.

Ferguson, B. 2004. City of Santa Barbara Desalination Presentation; Charles Meyer Desalination Facility.

California Regional Water Quality Control Board, Central Coast Region. 2019. Waste discharge requirements for the City of Santa Barbara El Estero Water Resource Center discharge to the Pacific Ocean (Order No. R3-2019-0046, NPDES No. CA0048143). California State Water Resources Control Board. https://www.waterboards.ca.gov/rwqcb3/board_decisions/adopted_orders/2019/elestero_permit.pdf (accessed November 13, 2025).

City of Santa Barbara. 2022. Desalination. https://santabarbaraca.gov/government/departments/public-works/water-resources/water-system/water-sources/desalination (accessed November 13, 2025).

Regional Water Quality Control Board, Central Coast Region. 2019. NPDES Permit No. CA0048143 for City of Santa Barbara El Estero Wastewater Treatment Facility. www.waterboards.ca.gov/centralcoast/board_decisions/adopted_orders/2019/elestero_permit.pdf (accessed November 13, 2025).

# CASE STUDY: SOREK SEAWATER DESALINATION PLANT

**Management Approach:** Off-shore intake, discharge to oceans

**Subcommittee Member:** Erin Messner

**Project Location:** Palmachim, Israel

**Desalination Process:** The water source for the Sorek Seawater Desalination Plant is the Mediterranean Sea. Pretreatment is a combination of conventional dual media gravity filtration (anthracite and sand) and MF by using cartridge filters. RO includes first pass followed by second pass (three-stage cascade). Posttreatment for water stabilization is accomplished by dissolving calcite with carbon dioxide and sulfuric acid and disinfection using sodium hypochlorite.

## WTP Information

- Capacity: 150 million $m^3$/year (109 mgd)
- Technology: RO
- Energy source: independent power producer
- Project type: build–operate–transfer for 25 years
- Footprint: 100,000 $m^2$ (24.7 acres)
- Commission Date: August 2013

## Background

The Sorek Seawater Desalination Plant is currently the largest plant within a massive desalination project in Israel that is aimed at addressing the country's shortage of potable water. The current seawater desalination amount in Israel is around 600 million $m^3$ per year (158 billion gal per year), which is 75 percent of the domestic water usage. During the next few years, this amount is expected to reach 900 million $m^3$ per year (237 billion gal per year). Sorek Desalination Plant is supplying 150 million $m^3$ per year (39 billion gal per year) for 1.5 million people. The plant started commercial operation in August 2013 and since then has supplied over 1.37 billion $m^3$ (361 billion gal) of clean water.

The Sorek Desalination Plant includes a unique design of 16-in. membranes that are loaded in vertical pressure vessels. After dealing with major challenges related to this design and to the manufacturing of such membranes, significant advantages arose, including a reduction of energy consumption, reduced maintenance, and improved equipment transportation. A reduction in capital expenditure cost also occurred, due to the reduced footprint of the project.

This plant is designed to produce potable water that must meet Israel's Ministry of Public Health regulations related to drinking water quality (see Table 8-1).

## Water Source

The source is open intake source from the Mediterranean Sea. Two intake heads are located 1.3 km (0.81 mi) and 2.4 km (0.62 mi) from the seashore to the intake pit. These intake pipes are cleaned periodically by a pigging procedure. The water temperature range is 16–32°C (60.8–89.6°F) and the salinity is 39–41 parts per million (ppm) of TDS. The offshore brine pipe length is 700 m (2,297 ft) longer than the intake pipes to ensure that the brine will not be pumped into the intake heads.

**Table 8-1 Israel drinking water quality targets**

| Substance | Measurement | Target |
|---|---|---|
| Total dissolved solids | ppm | ≤300 |
| Chlorides | ppm | ≤20 |
| Sodium | ppm | ≤30 |
| Total hardness | ppm as calcium carbonate | 80–120 |
| Alkalinity | ppm as calcium carbonate | >80 |
| Boron | ppm | ≤0.3 |
| pH | — | 7.5–8.5 |
| Langelier saturation index | — | 0.0–0.5 |
| Turbidity | ntu | ≤0.5 |

ppm — parts per million

## Pretreatment

Pretreatment includes two sequential steps: conventional gravity DMFs followed by cartridge filters. The water from the intake pumps is divided into two symmetrical plants, northern and southern. Each plant contains 24 DMFs that are backwashed periodically using DMF filtrate water. The backwash is treated in an effluent treatment plant to separate the water (which is sent back to the sea) from the sludge (which is sent for landfill disposal after a filter-press stage).

The DMFs' filtrate is pumped into 24 filter tanks (per plant), each containing 220 cartridge filters with cutoff of 20 mm. When the delta pressure reaches the threshold value, the filters are replaced with new ones. The replacement rate depends on the seawater quality and the DMFs' efficiency.

## Desalination

The desalination in Sorek is based on two passes: the first pass is SWRO and the second a BWRO cascade. The 16-inch–diameter membranes are loaded in vertical pressure vessels.

Before reaching the RO membranes, the micronic filtrate is divided into two streams; approximately 50 percent heads toward the high-pressure feed pumps and high-pressure pumps (HPPs), and the second half is sent to the ERD. The number of HPPs in operation can be two to five pumps per plant, according to the electricity rate and the production plan. There are 14 SWRO trains per plant (50 vertical pressure vessels, each containing eight membranes, for a total of 11,200 membranes). The ERD system includes 11 DWEER units per plant. The SWRO product is divided into front matter, which is sent directly to posttreatment, and rear matter, which is desalinated again in the second pass. The second pass includes a three-stage cascade (a total of 5,000 membranes) with a booster pump between the first and the second stage. The main purpose of the second pass is to reach the threshold value of 0.3 ppm boron; therefore, its feed pH is increased by using caustic soda. The RO trains are cleaned periodically with acidic or basic solutions to recover the membranes' performance.

## Posttreatment

The threshold values for water stabilization require an addition of calcium and bicarbonate. Approximately 50 percent of the RO filtrate is pumped into 14 calcite cells, following a pH reduction by carbon dioxide and sulfuric acid. The calcite cells are loaded by a crane and flushed with product water. The product is then mixed with the other half (which contains a low ion concentration) after a pH correction by caustic soda. The product is then disinfected by hypochlorite and supplied to Mekorot, the national water company in Israel, which distributes the water to domestic, agricultural, and industrial end-users.

# CASE STUDY: TAMPA BAY DESALINATION PLANT (TBWDP)

**Management Approach:** Colocation with power plant

**Subcommittee Members:** Zita Yu, Greg Wetterau

**Project Location:** Tampa, Fla.

**Desalination Process:** Tampa Bay Desalination Plant sources water from neighboring Big Bend Power Plant (BBPP). Sourced water follows a standard pretreatment process, including coagulation, flocculation, and sedimentation followed by media filtration. Water proceeds through first-pass RO with a partial second pass, followed by posttreatment disinfection and regional blending before customer distribution.

## WTP Information

- Initial capacity: 25 mgd (94 ML/d) (product water)
- Current capacity: 25 mgd (94 ML/d)
- Technology: RO
- Originally commissioned: 2003

## Abstract

The TBWDP, currently owned by Tampa Bay Water (TBW) and operated by US Water Services Corporation American Water-ACCIONA, is the first large-scale seawater desalination plant built in the United States through a public–private partnership. The history and evolution of the plant have provided lessons on the procurement process as well as the pretreatment design for many other desalination projects that follow. This case study summarizes those key lessons learned and provides an update on TBW's vision for the future of this plant.

## Project Overview

As shown in Figure 8-8, the TBWDP facility is colocated with the 1,200-megawatt (MW) BBPP. Originally designed to produce 66 mgd (250 ML/d) of desalinated water for potable use, the treatment plant currently produces up to 25 mgd (94 ML/d) of drinking water. The facility does not have its own seawater intake. Instead, raw water is sourced from the power plant intake. The plant is equipped with 0.5-mm mesh Ristroph-type traveling screens to reduce I&E of fish eggs and larvae (WateReuse 2011). Screened influent goes through conventional pretreatment processes: a coagulation/flocculation/sedimentation process, media filtration using sand filters followed by diatomaceous earth filters and cartridge filters, followed by first-pass RO with partial second pass. The RO process is equipped with energy recovery turbines to reduce energy consumption for the RO process. The posttreatment process for stabilization uses chemicals for disinfection of the desalinated water. The product water is then pumped to the high-service pump station for regional blending before distribution to the customers. The RO brine is discharged into the power plant's cooling stream and blended with up to 1.4 billion gal (5.3 billion L) of cooling water, achieving a blending ratio of up to 70:1 and attaining an average salinity level 1.0–1.5 percent higher than water from Tampa Bay. The blended water flows through a discharge canal to blend further with more raw seawater for greater dilution of the mixture, achieving a salinity to ambient level (Tampa Bay Water 2010).

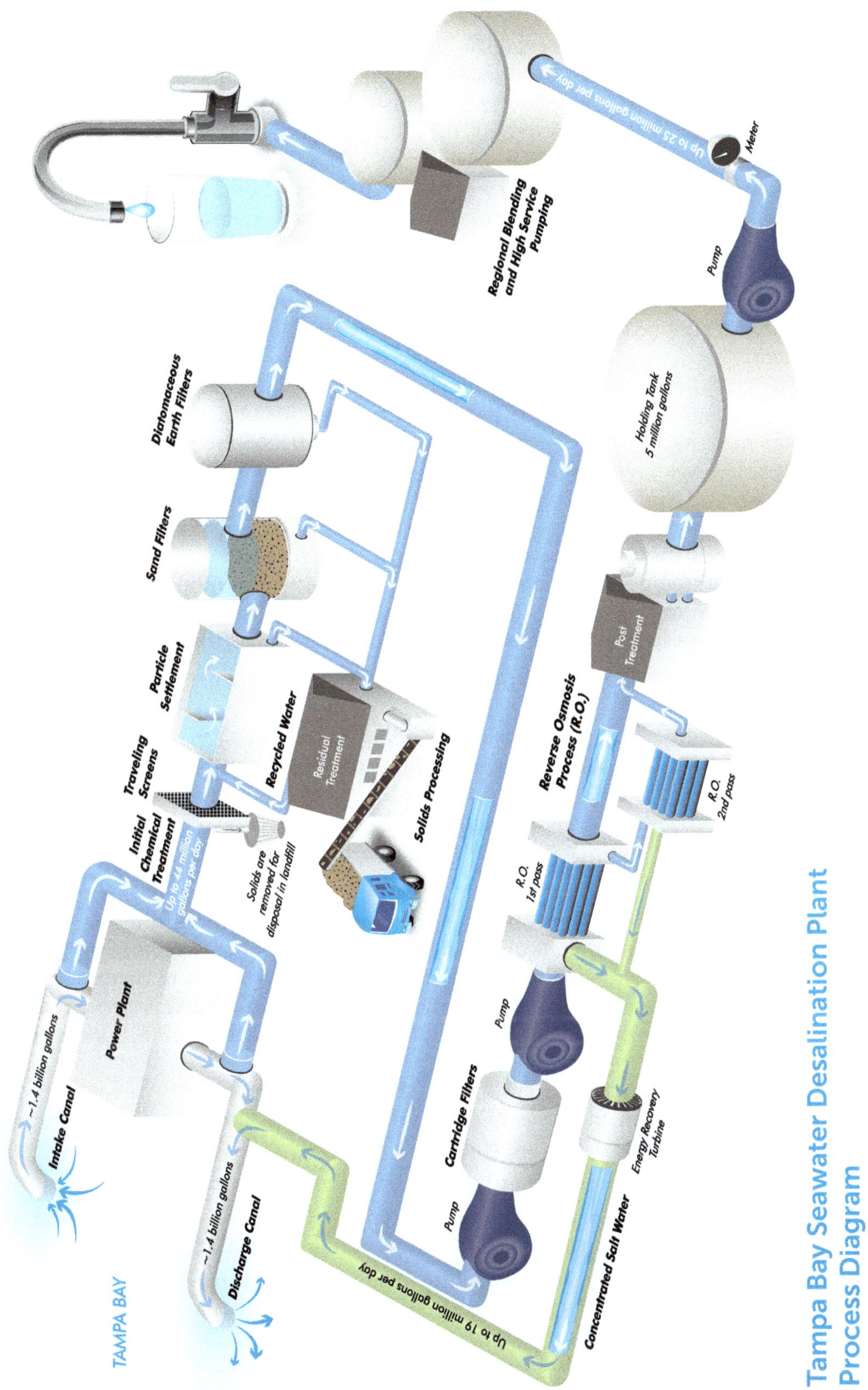


*Source:* Tampa Bay Water

**Figure 8-8 Schematic of the treatment system**

## Project Background

Formed in 1998, TBW is Florida's largest wholesale water supplier that provides water supply to Hillsborough, Pasco, and Pinellas counties and the cities of New Port Richey, St. Petersburg, and Tampa. The plant was designed to provide 10 percent of the region's demand, or 10 mgd (38 ML/d; Rand 2003), to help the region meet the requirement of reducing its groundwater use by 50 percent by 2008 (Jimenez et al. 2007). TBW opted for using a public–private partnership delivery model via design–build–own–operate–transfer (DBOOT) for the project to manage risk. TBW selected Poseidon Resources (now Poseidon Water) to be its private partner, with Poseidon providing up to at least 10 percent of the overall investment in the facility through a competitive process. Poseidon was responsible for obtaining permits, operating a pilot plant to validate the proposed process design, and bringing the plant online within the set schedule. Poseidon also assumed the risk for plant development.

At 50-percent construction completion, TBW acquired the plant from Poseidon to transfer the partnership from DBOOT to a traditional design–build–operate, with Covanta Water remaining as the engineering, procurement, and construction contractor and Poseidon serving as an advisor for a year. The plant started commissioning in 2003 but was unable to meet performance specification during the acceptance testing (Jimenez et al. 2007).

The plant went through significant modifications to address the treatment deficiencies, increase production, and enhance operability and reliability. Currently, the plant is capable of producing an average of 25 mgd (95 ML/day) of drinking water to the service area. The plant is operated intermittently to meet seasonal peak demands within the service area.

## Water Quality

The source water of the TBWDP is from Hillsborough Bay with TDS influenced by tides, rivers, and estuaries, fluctuating between 5 and 32 g/L (FDEP 2010). As a plant that uses surface intake via the BBPP's intake, additional screening to remove large debris was introduced to improve solids handling after the plant modification.

## Design Modifications

When the plant was originally designed, the pretreatment provided to the RO process included addition of ferric sulphate ahead of a two-stage sand filter to remove particulates and solids. Additionally, the media were continuously backwashed to help reduce the silt density index (SDI) of the RO feed water, and pH adjustment was also provided. However, soon after its construction, the original plant was unable to meet performance specifications.

Before commissioning a system overall, a series of evaluations was conducted to assess the potential causes and develop appropriate remedies, including the RO autopsy. The autopsy concluded that (Jimenez et al. 2007)

- membrane flux was dropped to 30 percent of the specifications;
- both organic and inorganic fouling were observed on the membranes, with the main inorganic foulants primarily iron, aluminum, and silica; and
- the polyamide active layers of the RO membranes were irreversibly damaged by oxidant levels.

The TBWDP went through design modifications to address pretreatment deficiencies. Numerous changes were made to the plant, including the use of variable-frequency drive pumps throughout, to maintain a stable operational conditions to accommodate the varying influent water quality (Harvey et al. 2020). The key changes are summarized in the following sections.

**Mechanical control for biofouling.** The TBWDP does not have its own seawater intake. Instead, raw water is sourced from the intake and discharge canals of the power plant. In the original design of the treatment facility, no separate screens were provided at the inlet of the treatment plant. During the design modifications, the inlet to the TBWDP is equipped with 0.5-mm mesh Ristroph-type traveling screens to reduce I&E of fish eggs and larvae (WateReuse 2011). The solids removed are disposed in landfills.

**Chemical control for biofouling.** One of the culprits for the performance challenges was the growth of invasive Asian green mussels in raw water pipes and filters. The use of sodium hypochlorite was determined ineffective in preventing growth of these mussels. Instead, chlorine dioxide was selected for biofouling control purposes (Jimenez et al. 2007).

**Coagulation-flocculation system.** A coagulation-flocculation system with separate coagulation and flocculation basins as well as a contact tank was added. Ferric chloride was selected as the main coagulant for the process (Jimenez et al. 2007).

**Sand filter modifications.** The two-stage sand filtration was changed to a single-stage upward-flow system This change enabled the filters to be operated at a lower filter flux by repurposing the second stage filters (Harvey et al. 2020).

**Additional media filtration.** The use of sand filters only in the original design was deemed insufficient for RO pretreatment. Precoat diatomaceous earth filters were added downstream of the sand filters to provide SDIs below 3 using particle counts, turbidity, residual iron, and SDI results as the basis for the optimization process. The optimal flux for the filters was 4.5 m/h (4,500 LMH or 2,646 gfd) (Jimenez et al. 2007).

**Solids handling.** A new solids-handling facility was also added to process solids collected from the contact basins and backwash of the media filters. The solids are thickened using coagulation, flocculation, and a lamella clarifier and dewatered using belt presses before disposal in a landfill. Water removed from the dewatering processes is directed back into the feed flow downstream of the traveling screens (Harvey et al. 2020).

**Posttreatment.** A lime system that consists of lime saturators, carbon dioxide addition, and a contactor was added to stabilize product water (Harvey et al. 2020; FDEP 2010). Sodium hypochlorite is used for disinfection, and sulfuric acid is used for pH adjustment.

## REFERENCES

FDEP (Florida Department of Environmental Protection) Division of Water Resource Management. 2010. *Desalination in Florida: Technology, Implementation, and Environmental Issues*. Tallahassee, FL: FDEP.

Harvey, N.J., Z. Rehman, T. Leiknes, N. Ghaffour, H. Urakawa, and T.M. Missimer. 2020. "Organic Compounds and Microbial Assessment of a Seawater Reverse Osmosis Facility at Tampa Bay Water, USA." *Desalination*, 496:114735.

Jimenez, A., G. Hijos, B.J. Moore, and J.J. Malfeito. 2007. Tampa Bay Seawater Desalination Plant: A Reliable Solution. International Desalination Association World Congress, Maspalomas, Spain.

Rand, H. 2003. "Tampa Bay Desalination Project—an Unfinished Story." *WaterWorld*. www.waterworld.com/home/article/16200492/tampa-bay-desalination-project-an-unfinished-story (accessed July 1, 2025).

Tampa Bay Water. 2010. Tampa Bay Seawater Desalination Plant. Project History, How the Plant Works Desal Fun Facts. https://www.tampabaywater.org/wp-content/uploads/Desalination-Fact-Sheet.pdf.

WateReuse. 2011. *Desalination Plant Intakes. Impingement and Entrainment Impacts and Solutions.* https://watereuse.org/wp-content/uploads/2015/10/IE_White_Paper.pdf (accessed July 1, 2025).

# Index

Note: *f* indicates figure; *t* indicates table

## R



## S

## T



## V



## W

# AWWA Manuals

M1, *Principles of Water Rates, Fees, and Charges*, #30001
M2, *Instrumentation and Control*, #30002
M3, *Safety Management for Water Utilities*, #30003
M4, *Water Fluoridation Principles and Practices*, #30004
M5, *Water Utility Management*, #30005
M6, *Water Meters—Selection, Installation, Testing, and Maintenance*, #30006
M7, *Problem Organisms in Water: Identification and Treatment*, #30007
M9, *Concrete Pressure Pipe*, #30009
M11, *Steel Pipe—A Guide for Design and Installation*, #30011
M12, *Simplified Procedures for Water Examination*, #30012
M14, *Backflow Prevention and Cross-Connection Control: Recommended Practices*, #30014
M17, *Fire Hydrants: Installation, Field Testing, and Maintenance*, #30017
M19, *Emergency Planning for Water and Wastewater Utilities*, #30019
M20, *Water Chlorination/Chloramination Practices and Principles*, #30020
M21, *Groundwater*, #30021
M22, *Sizing Water Service Lines and Meters*, #30022
M23, *PVC Pipe—Design and Installation*, #30023
M24, *Planning for the Distribution of Reclaimed Water*, #30024
M25, *Flexible-Membrane Covers and Linings for Potable-Water Reservoirs*, #30025
M27, *External Corrosion Control for Infrastructure Sustainability*, #30027
M28, *Rehabilitation of Water Mains*, #30028
M29, *Water Utility Capital Financing*, #30029
M30, *Precoat Filtration*, #30030
M31, *Distribution System Requirements for Fire Protection*, #30031
M32, *Computer Modeling of Water Distribution Systems*, #30032
M33, *Flowmeters in Water Supply*, #30033
M36, *Water Audits and Loss Control Programs*, #30036
M37, *Operational Control of Coagulation and Filtration Processes*, #30037
M38, *Electrodialysis and Electrodialysis Reversal*, #30038
M41, *Ductile-Iron Pipe and Fittings*, #30041
M42, *Steel Water-Storage Tanks*, #30042
M44, *Distribution Valves: Selection, Installation, Field Testing, and Maintenance*, #30044
M41, *Fiberglass Pipe Design*, #30045
M46, *Reverse Osmosis and Nanofiltration*, #30046
M47, *Capital Project Delivery*, #30047
M48, *Waterborne Pathogens*, #30048
M49, *Quarter-Turn Valves: Head Loss, Torque, and Cavitation Analysis*, #30049
M50, *Water Resources Planning*, #30050
M51, *Air Valves: Air-Release, Air/Vacuum and Combination*, #30051
M52, *Water Conservation Programs—A Planning Manual*, #30052
M53, *Microfiltration and Ultrafiltration Membranes for Drinking Water*, #30053
M54, *Developing Rates for Small Systems*, #30054
M55, *PE Pipe—Design and Installation*, #30055
M56, *Nitrification Prevention and Control in Drinking Water*, #30056
M57, *Algae: Source to Treatment*, #30057
M58, *Internal Corrosion Control in Water Distribution Systems*, #30058
M60, *Drought Preparedness and Response*, #30060
M61, *Desalination of Seawater*, #30061
M62, *Membrane Applications for Water Reuse*, #30062
M63, *Aquifer Storage and Recovery*, #30063
M65, *On-Site Generation of Hypochlorite*, #30065

M66, *Cylinder and Vane Actuators and Controls—Design and Installation,* #30066

M68, *Water Quality in Distribution Systems,* #30068

M69, *Inland Desalination and Concentrate Management,* #30069

M77, *Condition Assessment of Water Mains,* #30077

M80, *Biological Drinking Water Treatment* #30080

www.ingramcontent.com/pod-product-compliance
Ingram Content Group UK Ltd.
Pitfield, Milton Keynes, MK11 3LW, UK
UKHW052011041026
15590UKWH00008B/2278

9 781647 172602